UNDERGROUND ENGINEERING FOR SUSTAINABLE URBAN DEVELOPMENT

Committee on Underground Engineering
for Sustainable Development

Committee on Geological and Geotechnical Engineering

Board on Earth Sciences and Resources

Division on Earth and Life Studies

NATIONAL RESEARCH COUNCIL
OF THE NATIONAL ACADEMIES

THE NATIONAL ACADEMIES PRESS
Washington, D.C.
www.nap.edu

THE NATIONAL ACADEMIES PRESS **500 Fifth Street, NW** **Washington, DC 20001**

NOTICE: The project that is the subject of this report was approved by the Governing Board of the National Research Council, whose members are drawn from the councils of the National Academy of Sciences, the National Academy of Engineering, and the Institute of Medicine. The members of the committee responsible for the report were chosen for their special competences and with regard for appropriate balance.

This study was supported by the National Science Foundation under Grant No. CMMI-0946245. The opinions, findings, and conclusions or recommendations contained in this document are those of the authors and do not necessarily reflect the views of the National Science Foundation. Mention of trade names or commercial products does not constitute their endorsement by the U.S. government.

International Standard Book Number-13: 978-0-309-27824-9
International Standard Book Number-10: 0-309-27824-4

Library of Congress Control Number: 2013932381

Additional copies of this report are available for sale from the National Academies Press, 500 Fifth Street, NW, Keck 360, Washington, DC 20001; (800) 624-6242 or (202) 334-3313; http://www.nap.edu/ .

Cover: Cover design by Michele de la Menardiere. *Top left* is a screw conveyor (used to remove material excavated from a tunnel) of the world's largest tunnel boring machine being constructed for tunnelling beneath downtown Seattle. *Top right* is the inside of an underground utilidor in Amsterdam; courtesy of H. Admiraal. *Middle left* is a laser scanned image of a building foundation under construction; courtesy of Y. Hashash. *Bottom* is a train departing from a Washington, D.C. underground Metro station.

Printed in the United States of America.

THE NATIONAL ACADEMIES

Advisers to the Nation on Science, Engineering, and Medicine

The **National Academy of Sciences** is a private, nonprofit, self-perpetuating society of distinguished scholars engaged in scientific and engineering research, dedicated to the furtherance of science and technology and to their use for the general welfare. Upon the authority of the charter granted to it by the Congress in 1863, the Academy has a mandate that requires it to advise the federal government on scientific and technical matters. Dr. Ralph J. Cicerone is president of the National Academy of Sciences.

The **National Academy of Engineering** was established in 1964, under the charter of the National Academy of Sciences, as a parallel organization of outstanding engineers. It is autonomous in its administration and in the selection of its members, sharing with the National Academy of Sciences the responsibility for advising the federal government. The National Academy of Engineering also sponsors engineering programs aimed at meeting national needs, encourages education and research, and recognizes the superior achievements of engineers. Dr. Charles M. Vest is president of the National Academy of Engineering.

The **Institute of Medicine** was established in 1970 by the National Academy of Sciences to secure the services of eminent members of appropriate professions in the examination of policy matters pertaining to the health of the public. The Institute acts under the responsibility given to the National Academy of Sciences by its congressional charter to be an adviser to the federal government and, upon its own initiative, to identify issues of medical care, research, and education. Dr. Harvey V. Fineberg is president of the Institute of Medicine.

The **National Research Council** was organized by the National Academy of Sciences in 1916 to associate the broad community of science and technology with the Academy's purposes of furthering knowledge and advising the federal government. Functioning in accordance with general policies determined by the Academy, the Council has become the principal operating agency of both the National Academy of Sciences and the National Academy of Engineering in providing services to the government, the public, and the scientific and engineering communities. The Council is administered jointly by both Academies and the Institute of Medicine. Dr. Ralph J. Cicerone and Dr. Charles M. Vest are chair and vice chair, respectively, of the National Research Council.

www.national-academies.org

COMMITTEE ON UNDERGROUND GEOENGINEERING FOR SUSTAINABLE DEVELOPMENT

PAUL H. GILBERT (Chair), Parsons Brinckerhoff Inc. (retired), Edmonds, Washington
SAMUEL T. ARIARATNAM, Arizona State University, Tempe
NANCY RUTLEDGE CONNERY, Independent Consultant, Brunswick, Maine
GARY ENGLISH, Seattle Fire Department, Washington
CONRAD W. FELICE, HNTB Corporation, Bellevue, Washington
YOUSSEF HASHASH, University of Illinois at Urbana-Champaign, Illinois
CHRIS T. HENDRICKSON, Carnegie Mellon University, Pittsburgh, Pennsylvania
PRISCILLA P. NELSON, New Jersey Institute of Technology, West Orange
RAYMOND L. STERLING, Louisiana Technical University (retired), Minneapolis, Minnesota
GEORGE J. TAMARO, Mueser Rutledge Consulting Engineers, Bristol, Rhode Island
FULVIO TONON, University of Texas at Austin, Texas

National Research Council Staff

SAMMANTHA L. MAGSINO, Study Director
MIRSADA KARALIC-LONCAREVIC, Senior Program Associate
NICHOLAS D. ROGERS, Financial and Research Associate
COURTNEY R. GIBBS, Program Associate
JASON R. ORTEGO, Research Associate (until June 2012)
CHANDA T. IJAMES, Senior Program Assistant

COMMITTEE ON GEOLOGICAL AND GEOTECHNICAL ENGINEERING

EDWARD KAVAZANJIAN, JR (Chair), Arizona State University, Tempe
JOHN T. CHRISTIAN, Consulting Engineer, Burlington, Massachusetts
PATRICIA J. CULLIGAN, Columbia University, New York, New York
DEBORAH J. GOODINGS, George Mason University, Fairfax, Virginia
MURRAY W. HITZMAN, Colorado School of Mines, Golden
JAMES R. RICE, Harvard University, Cambridge, Massachusetts

National Research Council Staff

SAMMANTHA L. MAGSINO, Senior Program Officer
CHANDA T. IJAMES, Senior Program Assistant

Preface

Underground infrastructure presents unique challenges for engineers because usable underground space is limited in its extent and is not easily observed or accessible. The safety, health, and welfare of the public at large are among the civil engineer's primary concerns while designing, constructing, maintaining, and operating physical infrastructure, including underground infrastructure. Underground engineers must rely on the skills and expert knowledge of all members of an interdisciplinary team to carry out their respective professional obligations within their scopes, budgets, and schedules.

A concept has recently been making its way into infrastructure systems requirements to be satisfied by the engineer: sustainability. There are numerous definitions of sustainability, but this report refers to sustainability as the ability to obtain and use resources to meet current needs and improve standards of living without compromising the ability of those in the future to do the same. Sustainable urban development includes the selective use of materials and resources and consideration of cost effectiveness, functionality, safety, aesthetics, and longevity. The concept of sustainability changes the scale of many engineering projects. Engineering for sustainability means that engineers will need to move beyond traditional practice and consider their projects as part of a far larger physical and social system. They will need to think about the functionality and behaviors of their projects over long time periods—perhaps well beyond the project's service life. This is especially true of underground infrastructure, the impacts of which on society can be widespread and beneficial, but the failure of which can be devastating, and the remnants of which—post-useful service life—can affect society and the use of the underground for centuries into the future.

The committee was provided a detailed statement of task intended to define the role of underground engineering and works in sustainable urban development, as well as to provide direction for a future research track that supports such engineering. The broad and complex nature of the task necessitated only high-level consideration of its numerous points. The committee determined that simply responding one by one to each of the bulleted items in the statement of task would not fully respond to the intent of the task as described by the study sponsors. Instead, the committee tackled each bullet through discussions of the definition of sustainability, the evolution of underground use, potential contributions of the underground to sustainable urban development, health and safety in the underground, technological challenges of underground engineering, and research and training needed to increase capacity for underground engineering that supports sustainable development.

The direction of committee deliberations and the report were informed through multiple discussions with the study sponsors. Dr. Richard Fragaszy of the National Science Foundation provided numerous important insights regarding the concept of sustainability. Dr. Jonathon Porter of the Federal Highway Administration also spoke with the committee to describe his agency's expectations regarding the committee task, and answered the committee's questions with care. Committee deliberations and writings were also informed through excellent presentations during open sessions of committee meetings by Mr. Gordon Feller, Cisco Systems; Dr. Edward Garboczi, National Institute for Standards and Technology; Mr. Michael Grahek, Los Angeles Department of Water and Power; Mr. F. G. Wyman Jones, Los Angeles County Metropolitan Transit Authority; Mr. Richard Little, Keston Institute for Public Finance and Infrastructure Policy, University of Southern California; Dr. Harvey Parker, Harvey Parker and Associates, Inc; Mr. Kevin Peterson, Peterson Design; Dr. Helen Reeves, British Geological Survey; Mr. Henry A. Russel, Parsons Brinkerhoff, Inc.; Dr. Benedict Schwegler, Jr., Walt Disney Imagineering Research and Development; and Dr. Raymond Sterling, Louisiana Technical University. Numerous others also contributed to the committee process through less formal discussions with individual committee members and National Research Council (NRC) staff. Although there are too many to list here, the committee owes a debt to each of these people.

The committee is also grateful to the numerous NRC staff that provided direction, assistance in text development, and logistical and research support over the duration of the project. Their contributions to this process kept us moving forward, focused on the statement of task, well fed, and well informed. Our NRC study director, Sammantha Magsino, was particularly valuable to the committee in turning the many original text drafts on a range of topics from each committee member into coherent and consistent sections, chapters, and finally the report.

The study process has made it clear to the committee that the underground engineering needed to develop urban sustainability will require engineers in professional practice to rethink how they have traditionally delivered their work products. It will also require those in research and education to consider new multidisciplinary approaches to improving technologies and increasing capacities. Engineering the underground permanently changes the underground—a valuable, and irreplaceable resource. It is the ethical responsibility of all making those changes to anticipate and understand the impacts of those changes to the larger physical and social infrastructures over time to avoid harming future generations, and, in fact, to help those future generations to thrive.

Adding to or changing the systems of systems that comprise urban infrastructure will demand that underground engineers become more multidisciplinary in their approaches and that they more comprehensively communicate and rely on the expertise of engineering scientists, planners, architects, and other professionals from all contributing disciplines. This report presents a foundation for how this professional transition can be made, and it presents a framework for new education, training, and research strategies to prepare engineers and all their colleagues for the future.

Paul H. Gilbert, P.E., NAE
Chair

Acknowledgment of Reviewers

This report has been reviewed in draft form by persons chosen for their diverse perspectives and technical expertise in accordance with procedures approved by the National Research Council's Report Review Committee. The purpose of the independent review is to provide candid and critical comments that will assist the institution in making its published report as sound as possible and to ensure that the report meets institutional standards of objectivity, evidence, and responsiveness to the study charge. The review comments and draft manuscript remain confidential to protect the integrity of the deliberative process. We thank the following individuals for their participation in the review of this report:

Arthur Bendelius, A & G Consultants Inc., Fayetteville, Georgia
Brenda Bolhke, Independent Consultant, Great Falls, Virginia
Joseph P. Colaco, CBM Engineers, Houston, Texas
Louise K. Comfort, University of Pittsburgh, Pennsylvania
Herbert H. Einstein, Massachusetts Institute of Technology, Cambridge
Derek Elsworth, Pennsylvania State University, University Park
Marc Pisano, University of Southern California, Los Angeles

Although the reviewers listed above have provided many constructive comments and suggestions, they were not asked to endorse the conclusions or recommendations, nor did they see the final draft of the report before its release. The review of this report was overseen by Charles Fairhurst, Itasca Consulting Group, Inc., Minneapolis, Minnesota. Appointed by the National Research Council, he was responsible for making certain that an independent examination of this report was carried out in accordance with institutional procedures and that all review comments were carefully considered. Responsibility for the final content of this report rests entirely with the authoring committee and the institution.

Contents

APPENDIXES

Summary

For thousands of years, the underground has provided refuge, resources, foundations for surface structures, and a place for spiritual or artistic expression. More recently, important infrastructure has been placed underground because of proximity to services, to preserve surface space, provide climate or security isolation and containment, reduce construction and energy costs, improve traffic flow, and for various aesthetic benefits. Underground space can provide three-dimensional freedom often unavailable in densely developed areas. Infrastructure systems can be placed beneath cities, under rivers, and even through mountains. Millions of people rely on these systems with little thought to the comfort and conveniences provided. Placing new infrastructure underground also may encourage or support the redirection of urban development into sustainable patterns. Resilient, well-maintained, and well-performing underground infrastructure, therefore, becomes an essential part of sustainability.

At the request of the National Science Foundation (NSF), the National Research Council (NRC) conducted a study to summarize current underground engineering knowledge, identify needed research and direction for a new research track to support sustainable development through underground engineering, and examine drivers that promote or inhibit underground development (see Box S.1 for statement of task). The NRC convened a panel including researchers and practitioners with expertise in geotechnical engineering, underground design and construction, trenchless technologies, risk assessment, visualization techniques for geotechnical applications, sustainable infrastructure development, lifecycle assessment, infrastructure policy and planning, and fire prevention, safety, and ventilation in the underground. The committee's report is intended to inform

BOX S.1
Statement of Task

An ad hoc committee of the National Academies will conduct a study to explore the potential advantages of underground development in the urban environment, to identify the research needed to take advantage of these opportunities, and to develop an enhanced public and technical community understanding of the role of engineering of underground space in the sustainability of the urban built environment, specifically the minimization of consumption of nonrenewable energy resources, construction materials, and negative impact on the natural, built, and social environments. In particular the study will:

- Summarize current geological and geotechnical engineering knowledge about underground development in the urban environment and how utilization of underground could increase sustainability, including knowledge of geologic site characterization, construction and geotechnical monitoring techniques, energy requirements, use of excavated materials, and lifecycle costs and benefits of underground infrastructure development.
- Identify the research needed to capitalize on opportunities for enhancing sustainable urban development through underground engineering, in the following areas:
    - Underground characterization, prediction of the geologic environment, and ground response critical for successful design and construction of underground projects and critical facilities to maximize sustainability and resiliency;
    - Construction and monitoring methodologies and enhanced excava-

public- and private-sector audiences engaged in research, urban and facility planning and design, underground construction, and safety and security.

Based on discussions with study sponsors, this report focuses on contributions of engineered underground space to sustainable development and outlines needs in the research, educational, regulatory, and social environments that would maximize those contributions. The report provides a set of overarching observations, conclusions, potential actions, and research topics related to integrated and interdisciplinary infrastructure systems design and management; underground engineering education, training, research, and practice; approaches to management and technological development; infrastructure lifecycle assessment; underground space use acceptance and safety; and underground space as a resource. These conclusions address all aspects of the charge generally rather than specifically. Important research topics are highlighted with the conclusions, but more are found throughout the main body of the report.

tion methods, including tunneling, conducive to sustainable and resilient underground development;

- Smart underground structures and conduits that report their status;
- Health and safety considerations, such as cost-effective ventilation, light, and concerns related to radon exposure or fire control;
- Lifecycle cost and benefit issues, including reduced energy needs for heating and cooling, reduced construction material use, use of excavated materials, increased longevity of underground structures and reduced maintenance associated with stable temperatures and isolation from surface weathering effects;
- The potential sustainability benefits of increased use of underground space for human transportation systems, including roadways and mass transit, and freight;
- The potential for integrating of energy, water, and waste systems for certain urban regions to improve sustainability; and
- How underground development might address concerns related to the impacts of climate change on the urban environment.

The committee will recommend directions for a new underground engineering research track focused on earth systems engineering and management to ensure future human resources for sustainable underground development, will analyze the advantages and disadvantages of establishing a new research center in this area, and consider other potential options for enhancing the human resource capacity for sustainable underground development (including the status quo). The committee also will consider from a social science point of view, the policy, economic, and human behavioral drivers that promote or inhibit the development of the subsurface in a sustainable manner, but will not make policy or funding recommendations.

THE UNDERGROUND FOR SUSTAINABILITY

Sustainability is defined in this report as the ability to meet present societal needs without compromising the ability of future generations to do the same. Maintaining or improving quality of life and maintaining long-term ecological balance are among societal needs. An unhealthful natural environment can negatively impact food, water, and air supplies and degrade quality of life and health to unacceptable levels. Resilience, an important aspect of sustainability, is defined as the ability to respond to environmental changes—especially natural or human-caused adverse events—with minimum impact on functioning.

Master plans of some cities (e.g., Singapore) include extensive underground use. Well-planned underground infrastructure can positively influence land use and development decisions and can reduce vehicle use and associated impacts. High-density urban centers may depend on centralized services but can capitalize on centralization to increase sustainability. Underground transportation infrastructure (e.g., urban roads and highways, public transit subways, grade-separated and underground freight railroads, high speed rail, and pedestrian rights of way) can address multiple growth-related challenges in urban areas (e.g., congestion, urban sprawl) if infrastructure elements are optimally designed and located. Well-planned and operated underground infrastructure can, in many cases, improve quality of life and sustainability more so than can similar-purpose surface infrastructure.

UNDERGROUND INFRASTRUCTURE AS PART OF A SYSTEM

Observation: There is little strategic coordination of underground infrastructure development in the United States.

Conclusion 1. Coordinated formal administrative support and management of underground infrastructure as part of an integrated, multi-dimensional, above- and belowground system of urban systems is vital to urban sustainability.

Potential actions:

a. Recognize responsibilities related to formal support for underground infrastructure as part of the total urban system through coordinated planning and operations, fostered technological development, and local and regional rule making.

b. Develop and encourage use of a system for consistent data collection, archiving, and access to be used by all facility owners and operators to aid decision making.

Research:

a. Explore within the federal government the most appropriate technical and administrative approaches to facilitate coordinated management of the underground as part of a total urban system. Recognize and coordinate with ongoing research in this area, for example, that conducted by the NRC Transportation Research Board related to road projects.

b. Conduct a technology scan of how countries and cities around the world collect, manage, make available, and use three-dimensional geological and buried structure information.

Infrastructure development, operation, and maintenance require management of the complex physical, social, and environmental systems influencing proper functioning. Development of underground infrastructure suffers in the United States from the lack of a mission agency or organization within the federal establishment dedicated to coordinating development activities across sectors. Project and research funding mechanisms tend to focus on solving particular problems with little consideration of long-term impacts on the total urban system.

Coordination could lead to better management of research investments, optimized decision making, reduced risk for federal development projects, and better leveraging with state and local entities. Although some planning and zoning by local governments of outward and upward city growth does occur, there is little analogous control of underground space, and even less control that coordinates above- and belowground development. It is possible and desirable to identify, protect, and effectively zone prime subsurface resources for optimal use as is done, for example, in Helsinki, Finland, Montreal, Canada, and Singapore. Some policy changes can result in lower costs through, for example, streamlining time-consuming permitting processes as is done in Japan.

Observation: Market forces in the United States encourage workforce capacity growth and urban and infrastructure development, but often in an ad hoc manner that may not be consistent with urban sustainability.

Conclusion 2. Development of underground space as part of sustainable urbanization requires expanded and coordinated communication with stakeholders to better incorporate site-specific conditions, greater flexibility, and long-term community needs into infrastructure system design and optimal lifecycle management.

Potential actions:

a. Establish a federally led interdisciplinary network or organization of organizations and institutions to guide sustainable patterns in underground infrastructure development and encourage interdisciplinary research and communication of findings among all disciplines and stakeholders. Stakeholders include, for example, designers, long-term planners, architects, safety specialists, and an array of engineering, geologic, geophysical, environmental, and contracting specialists from industry, government, and academia.

b. Develop mechanisms for integrated and holistic three-dimensional research and planning that include information management and communication technologies to facilitate complex research, design, construction, operation, and management of underground infrastructure.

Research:

a. Explore models for designing sustainability into engineered systems of urban systems that recognize interdependencies, vulnerabilities, complexity, and adaptability. Coordinate ongoing research in the United States and elsewhere on, for example, complex adaptive systems and human factors engineering (e.g., incorporating behavioral science, human performance and capacity, personnel and training, and human biology and physiology into engineered systems).

b. Develop conceptual models of the complex interactions among multiple systems (e.g., mechanical, human, and environmental) to improve understanding, reduce risk, and effectively manage infrastructure amid changing technologies, societal conditions, and expectations.

c. Research the behavior of those operating, maintaining, and using underground infrastructure during normal and worst-case operation scenarios to optimize the human-technical interfaces in a manner consistent with long-term values.

Underground infrastructure development is a multidisciplinary endeavor. A sustainable urban system is possible if decisions are informed by the links between the social, technical, and governing elements of society (as occurs to some extent today). Underground infrastructure projects, however, are often undertaken on a project-by-project basis with minimal consideration of long-term maintenance or societal needs. This approach is inconsistent with sustainability. Decisions are often made among decision makers with competing political, social, and economic interests and security concerns that influence if, how quickly, where, and by what methods underground development occurs. To maximize sustainability, multidisciplinary efforts are needed during the entire infrastructure life cycle.

Better informed decision making is possible when engineers understand the complex and interactive social and economic factors that contribute to sustainability and when urban planners have realistic expectations about the underground environment. Some interdependencies are obvious, but other interdependencies—some critically important to national security—may remain unknown without appropriate communication and planning among experts and stakeholders.

The capacity for flexibility is needed to address emerging issues, technologies, and societal expectations during and beyond the operational life of underground infrastructure. New hazards associated with vulnerable and deteriorating infrastructure systems, climate change, and security concerns, for example, may affect provision of service, environmental quality, or personal safety. Extreme events (e.g., terrorist acts or natural disasters) present still other hazards and risks. Sustainability depends on planners and engineers building and pooling capacities to anticipate and accommodate human behavior and the constantly evolving urban environment. Accounting for human behavior in underground space can lead to creation of environments that allow more intuitive understanding of safety in the underground under varied circumstances.

A new institutional framework for professional planning, architecture, engineering, public administration, and social and economic policy committed to sustainable development will be difficult to create but could recharge U.S. educational and research capacities to address sustainable urban underground space use. Federal, state, and local agencies, the engineering and construction industries, and university educators and researchers need to work together to increase human capital.

Observation: Complex ownership models for underground infrastructure confuse responsibility for routine inspections, maintenance, repairs, guidelines, budgets, and liability.

Conclusion 3. There is a need to understand the ownership and control models of underground space and to develop guidelines for funding and performing essential periodic inspections, maintenance, and repair of individual infrastructure elements.

Research:

a. Analyze multidisciplinary and holistic approaches to view the complex web of ownership, control, and responsibilities associated with maintenance and safety of underground infrastructure.

b. Examine multidisciplinary approaches to aid transition to more modern systems management.

Underground infrastructure in the United States is typically owned and controlled by numerous individuals, partnerships, corporations, and local, state, and federal government. Responsibility for routine inspections, maintenance, and repairs is confused, and ambiguity regarding applicable guidelines, budgets, and terms can arise. Separate agencies deal independently with transportation, housing and urban development, homeland security, and energy issues. Sustainability goals will be hindered without more coordinated control and management.

STATUS OF U.S. RESEARCH, TECHNOLOGY DEVELOPMENT, AND EDUCATION

Observation: The United States was a world leader in many areas of underground science and technology when there was federal and industry investment in underground engineering research and development.

Conclusion 4. Maintaining global competitiveness in underground engineering education, technology development, and practice supports urban sustainability, resilience, and the standard of living of the United States.

Potential Action:

Allocate resources for broader interdisciplinary education and technology development in underground design and construction.

Research:

Expand U.S. research that advances and revolutionizes, for example, materials technologies, robotic construction technologies, laser guidance systems, geographic information systems, and enhanced computer analysis and visualization systems that improve the ability to model, design, plan, and reduce risk associated with complex underground systems (see Chapter 6 for more detail).

Geotechnical expertise will always be critical to delivery of underground facilities with lower costs and risk and to better lifecycle performance. Geotechnology education, research, and practice need to better integrate all disciplines related to site investigation, design, construction, operation, and risk management of underground facilities. The complexity and unpredictability in underground construction indicate that many challenges remain. Technological advances improve our ability to understand, model, construct, and reduce risk associated with underground infrastructure. It is not in the country's best interest, however, to rely heavily on imported technological advances and expertise to create and maintain underground facilities, as has become a trend in the United States. Much new knowledge, technology, and project-specific memory may leave the country at the completion of construction, to the possible long-term detriment of underground infrastructure operation, maintenance, and security.

Observation: Lack of funding continuity that allows meaningful investment in equipment and faculty has resulted in a substantial reduction in the number of U.S. university programs dedicated to integrated underground engineering research and education.

Conclusion 5. There is a critical shortage in educational, training, and research opportunities for engineers who wish to learn and practice underground engineering in the United States.

Potential actions:

a. Develop national multidisciplinary, multi-institutional, cross-sector research centers that focus on different areas in underground engineering and sustainable urban infrastructure to produce the next generation of leaders in underground engineering.

b. Integrate graduate underground engineering studies with research programs or a critical mass of coordinated faculty activity to anchor research to existing programs. Create opportunities to specialize in particular aspects of underground engineering, but with a multidisciplinary approach.

c. Develop university consortia to aggregate faculty expertise; strengthen industry-university faculty relationships.

d. Teach better facility planning and management with a multidisciplinary approach through traditional, distance, or hybrid-style education formats. Traineeships (e.g., NSF's Integrative Graduate Education and Research Traineeships) could help to fund programs.

e. Expose undergraduates to multiple disciplines, issues, challenges, and opportunities associated with sustainable underground space use and engineering.

f. Develop continuing education opportunities for professionals.

g. Develop appropriate credentialing for inspectors, technicians, and operators of complex underground facilities.

Underground engineering knowledge, expertise, and training in the United States today are obtained mostly through mentoring and on the job experience, rather than through higher education. Such training provides hands-on experience and benefits the workforce, but competitiveness and liability concerns can limit information sharing more generally within the industry and can limit exposure for young engineers to a range of technologies and methodologies. Because few commercial incentives exist for industry to embrace the challenges associated with long-term infrastructure or urban system sustainability, young engineers may not be exposed to potential solutions for these issues. In contrast, U.S. students educated within multidisciplinary U.S. research institutes are more likely to benefit from the advances and broad knowledge and technology dissemination that takes place and become a better prepared domestic workforce.

Optimized design and more judicious use of resources can result from detailed knowledge of the underlying and nearby geology and human-development histories (e.g., existing infrastructure and legacy construction materials) and the ability to minimize unanticipated ground conditions. Traditional undergraduate programs do not teach an integrated approach to practice, and few graduate programs offer interdisciplinary programs in underground engineering, certification in specific areas (e.g., tunneling), or specialization within more general graduate degree programs that allow for optimization. Knowledge of technologies for tunneling (including trenchless), excavation, ground support, ground improvement, and natural and built systems monitoring, and other functions is essential. But good education programs also will include mechanical, electrical, civil, structural, geotechnical and geological engineering; planning; architecture; public policy; fire safety; and information technology in their curricula.

Few U.S. university faculty research tunnel design and construction performance. The lack of a continuous government focus on infrastructure issues, and the fragmentation of U.S. government-sponsored underground development research across several disciplines at the core of underground engineering (e.g., structural, geotechnical, and mining engineering), result in little expectation of program funding continuity. Opportunities in specialized areas such as tunneling are disappearing as a result of mandated reductions in credit hour requirements for undergraduates and a recent lack of interest by U.S. students in pursuing advanced degrees.

IMPROVING PERFORMANCE

Observation: The complexity of urban infrastructure systems and uncertainties associated with system design and performance increase with greater and more varied demands on both above- and belowground infrastructure.

Conclusion 6. Engineers and urban planners could better improve whole lifecycle facility performance and overall urban sustainability with documented and validated risk-informed approaches to project planning and design that balance lifecycle project needs in terms of service delivery, initial costs, resilience against extreme events, and effective maintenance and operations.

Research:

a. Advance existing and develop new technologies for modeling uncertainty during all phases of infrastructure life cycle. These include invasive and noninvasive technologies for geologic site characterization (including existing and legacy infrastructure and materials); analytical and computational design methods; excavation, ground support, and monitoring technologies; and technologies for asset management including related to the management of data and security (see Chapters 6 and 7 for more details).

b. Develop strategies to investigate potential hazards, impending problems, and cascading evolution of problems, especially given current underinvestment in infrastructure system rehabilitation.

c. Engineers and planners could use extreme events to understand complex systems behaviors and interdependencies and to validate computational models of system performance.

Full assessment of lifecycle costs and benefits may convince owners and planners that greater initial investment in underground infrastructure can be economical in the long term. Security and resilience of urban areas can be enhanced if decisions are informed by complete evaluation of the merits, deficiencies, and interactions of infrastructure elements with respect to all potential hazards and risks. Long-term performance of infrastructure can be improved with access to good models and data for analyses. However, the validity of models developed for individual system functionality and performance is often questionable, and uncertainty increases when modeling systems of greater complexity. Models of integrated systems of systems such as urban infrastructure have yet to be developed and validated.

LONG-TERM COMMITMENT

Observation: Aging underground infrastructure may be susceptible to deterioration and issues associated with changing technologies, changing climate, and societal needs.

Conclusion 7. Underground space development requires a long-term commitment to technological advancements in an environment that is friendly to improved planning, innovation, and implementation.

Potential actions:

a. Design infrastructure that allows ease of access for inspections, maintenance, repairs, upgrades, and reconfigurations in response to new needs or technologies that allow such work to be completed at lower costs.

b. Consider resource needs, availabilities, and access when making administrative and technical decisions concerning development. These include energy resources (e.g., oil, gas, and other energy resources), industrial minerals, high-value or critical strategic minerals (e.g., gold, uranium, rare earth elements), and construction materials (e.g., gravel, sand, building stone).

c. Use appropriate models that demonstrate multiple potential scenarios and allow better infrastructural system planning based on local conditions.

Research:

a. Academia and system stakeholders could collaboratively develop long-term performance simulation models for complex systems and validate the results over time to understand dynamic responses and emerging system behaviors.

b. Explore how technologies and innovations from other industries (e.g., exploration tools, in situ analytical techniques, measurement-while-drilling systems, laser scanning, fusion of multi-sensor data) and civilian application of military research could be applied to underground engineering.

c. Conduct long-term research on the effects of the underground infrastructure on the natural and built environments to increase the capacity of decision making for society's best long-term interests.

d. Research comprehensively and on a common risk-cost-reward basis the long-term effects on sustainability of underground storage or disposal of urban wastes (e.g., municipal, sewage, or energy-related products).

Lifecycle planning aids long-term infrastructure health. Age, deterioration, and changes in technologies and use mean that underground infrastructure systems constantly require attention. Selecting the most sustainable approaches to underground space use may be more likely if the best available science, technology, and ideas can evolve, keep up with societal needs, and become less expensive to use. For example, combining utility services into common utility tunnels (called utilidors) can isolate utilities from the surface in a continuously accessible location. Tangles of utility infrastructure in many urban areas can be reduced or avoided (such infrastructure typically remains in place long after its operational life), and more of the subsurface can remain available for other uses. This is particularly beneficial in areas with narrow rights-of-way, and can be economically advantageous when cost considerations include the value of the underground.

Strategic construction and long-term maintenance of underground infrastructure may result in fewer adverse environmental impacts than for surface infrastructure. Technological advances can help minimize noise and vibrations, protect air quality, and allow for recycling and reuse of waste construction materials, including soil and rock from a site. Technological advances that allow better prediction of impacts to water quality, groundwater flow, soil geochemistry, and underground temperatures and heat flow that may impact the natural and built environments are needed.

LIFECYCLE ASSESSMENT

Observation: Few data exist regarding the environmental and social impacts and lifecycle sustainability of urban development that can inform technology and administrative decisions related to long-term (decades to centuries) infrastructure operation, maintenance, and reduced costs.

Conclusion 8. Comprehensive and scientific retrospective studies of the direct and indirect costs and impacts of various types of underground projects are needed to evaluate usefulness and economic, environmental, and social impacts so that future planning can maximize sustainability.

Research:

a. Conduct comprehensive and scientific investigations to retrospectively identify the lifecycle performance of various types of underground infrastructure and to identify the aspects of project planning, design, construction, and operation that contribute most to project costs and performance. For example, track financial (both direct and indirect), environmental (e.g., air and water quality), and social impacts over an extended period (e.g., decades) following a project such as Boston's Central Artery alignment.

b. Develop common metrics for assessing sustainable development more generally, and for assessing specific economic, environmental, and social impacts.

c. Develop quantitative methods to compare the value of underground space on a par with other urban resources (e.g., linked to market value of surface property) and in consideration of the impacts on future underground use (e.g., infrastructure may need to be placed in increasingly difficult ground conditions).

d. Compile data about sustainability aspects of various construction methods and materials (e.g., the availability of materials and energy embodied in production of materials).

Planning horizons for decision makers are often far shorter than the useful life of underground infrastructure. Underground infrastructure development may require seemingly cost-prohibitive initial investment for construction when compared to similar-use surface infrastructure. Few data exist to validate investment support when long-term benefits are not valued. Lifecycle assessment can provide data through consideration of costs, impacts, and benefits—from raw materials acquisition, to construction and operations, through closure, decommissioning, and post-operational use. Additional inputs such as energy (e.g., for lighting and ventilation) also are factors. Similarly, understanding how some underground development has precluded or made other uses of underground space more expensive may inform decisions that affect future options. The costs and challenges of re-using occupied underground space remain long-term issues.

USER ACCEPTANCE, SAFETY, AND COMFORT

Observation: Underground infrastructure can safely enhance the lives of millions, but few federal-level safety regulations exist to guide operational safety at a time when underground system complexity is increasing.

Conclusion 9. Greater user acceptance and occupancy of underground infrastructure and facilities are likely if underground spaces are planned with more consideration of utility, ease of access, wayfinding, safety, and aesthetics.

Potential actions:

a. Develop and adopt performance-based safety mechanisms and codes that not only account for today's underground occupancies (e.g., mixed use, multi level) and risks, but also allow for expansion and change of use. The International Code Council technical requirements, applicable National Fire Protection Association standards, and other related standards and guidelines could be expanded and made applicable to underground facilities.

b. Incorporate human factor and complex systems engineering concepts to guide threat recognition and technical and operational decision making for normal operations and for operations during times of stress (e.g., in response to extreme events).

c. Incorporate behavioral science, training, biology and physiology, human performance and capacity into safety codes and design.

Research:

a. Research the state of practice and best practices related to safety systems (e.g., hazard detection, notification, ventilation, fire suppression, emergency egress, and system integration). Develop appropriate minimum safety system requirements to incorporate into national-level guidelines and standards.

b. Compare international underground safety codes and guidelines with those applicable in the United States to identify inadequacies and guide future practice, recognizing existing efforts in this area (e.g., by FHWA).

Underground space can be as safe, attractive, stimulating, functional, productive, and healthy as similar-use surface space. Negative perceptions about underground space, however, can be as difficult to overcome as complex safety and technical challenges. Acceptance and use of underground space may increase with greater convenience and comfort of use (e.g., by incorporating better connectivity among underground systems that limit pedestrian travel time and lengthy vertical movements by stairs, escalators, or elevators). More intuitive understanding of safety in the underground by its occupants will also increase acceptance.

Safety in the underground is achieved by avoiding, transferring, or reducing risks associated with naturally occurring phenomena (e.g., gases, radiation, extreme temperatures, water, and lack of oxygen) and human activity (e.g., fire, smoke, hazardous materials, intentional or accidental explosions, structural failure, or simple human failure). Safety is more challenging with increasing infrastructure complexity. Human factors engineering becomes essential to increasing the ability of people to operate and occupy the underground safely.

Safety codes are often written in response to incidents or litigation and are not flexible enough to accommodate evolving technologies. Safety is created operationally or through technical solutions, but it is dependent on designing and operating beyond mere compliance with often inadequate codes. The few federal-level safety regulations for underground infrastructure mostly apply to construction rather than to operational usage of most facilities. State-level fire safety codes do not fully address underground structures and will likely be inadequate when different occupancy types are combined in one underground space (e.g., public transportation and commercial).

Capital construction and operational risk mitigation costs for underground space can be substantial and could preclude an underground project from being started, or could result in improper system maintenance. Human factors engineering can help to minimize costs associated with avoiding or transferring risk, for example, by identifying ways to reduce risk through safety regulations and education when technological solutions are not feasible. Innovation in design and construction is fostered by moving beyond prescriptive and potentially ineffective codes toward performance-based mechanisms.

THE UNDERGROUND AS A RESOURCE

Observation: Underground space is a valuable but decidedly nonrenewable resource.

Conclusion 10. Underground space can enhance urban sustainability only if the underground is thoroughly understood and if underground use and reuse and the protection of the natural and built environments are incorporated into long-term total urban infrastructure system planning.

Potential actions:

a. Institute planning of all underground space as part of an evolving urban system to be carefully engineered or preserved for optimal long-term use and regional sustainability.

b. Establish reasonably intensive groundwater, soil, and infrastructure monitoring practices to track the health of the underground urban environment according to the general geologic conditions and use. Use data generated from a range of environments and situations to inform urban planning in other areas.

The underground is not a universal alternative to the surface, but many uses of underground space contribute to urban sustainability. It is critical that policies and administrative structures provide appropriate and comprehensive guidance, that the public develops a long-term community vision, and that community expectations regarding underground services are informed and met. An adequate institutional commitment to enhancing interdisciplinary and cross-sector research, education, and training capacity is needed to ensure the nation develops the types of underground infrastructure that support sustainable urban development economically, securely, and in a manner consistent with national priorities.

1

Introduction

In 2000, the National Academy of Engineering published a list of the 20 engineering achievements in the 20th century that included electrification, the automobile, water supply and distribution, computers, telephone, air conditioning and refrigeration, highways, the Internet, petrochemical mechanization, laser and fiber optics, nuclear technologies, and high performance materials (NAE, 2000). Many of these achievements have been described as mainstays of contemporary urban life (Papay, 2002), and many of the essential services linked to them are delivered using the urban underground during some stage of production, storage, and distribution. Maintenance and improvement of those services, as well as of the quality of life in urban regions, depend on a steady stream of investment and technological innovation.

Human activity and population growth, however, are transforming the nation and planet. Long-term challenges for society include learning how humans can prosper without continued degradation of Earth (Kammen and Jacobson, 2006) and how to make suitable and sustainable adaptations. Improving or even sustaining current standards of living in the future will place more stress on earth systems, especially in urban environments where population increases are expected. Approximately 80 percent of people living in the United States live in urban areas (U.S. Census Bureau, 2011). Approximately 53 percent of the American population lives within 50 miles of a coast (Markham, 2008) at a time when global climate change is predicted to have significant coastal impacts including sea level rise, changes in weather patterns (e.g., IPCC, 2007), and degradation of drinking water supplies (IPCC, 2008). Meanwhile, some suggest short-term focus needs to be on design and adoption of community-based strategies to reduce vulnerability to the potentially destructive impacts of climate change throughout the nation (NRC, 2010).

Intensive and well-coordinated use of underground space may be a key component of the sustainability solution. Engineers of underground space will have a vital role in planning, designing, constructing, operating, maintaining, and regulating underground space as well as in informing the social, economic, and even political decisions related to underground space and urban development. Increased interest in underground construction and development is evident throughout the world (Sterling and Godard, 2000). Underground engineering can provide a means to reduce energy use, increase green space preservation, sustainably process and store water and wastes, securely and efficiently site critical infrastructure, prevent and reverse degradation of the urban environment, and enhance quality of life. Many urban areas already enjoy the benefits of using underground space. The I-93 Central Artery and the I-90 extension in Boston (known collectively as the "Big Dig"), for example, although expensive, controversial, and not without problems, have improved peak period travel times through downtown Boston, saving an estimated $168 million in annual downtown travelers' costs and time (Massachusetts Turnpike Authority, 2006), and have resulted in an enhanced downtown cityscape. Sweden's experience with underground sewage treatment facilities since the 1940s (Isgård, 1975) and Norway's expansive network of underground infrastructure, including electric power generation, water supply and wastewater treatment facilities, air traffic control, financial, archival, civil defense and national security facilities (Linger et al., 2002) demonstrate that underground facilities can be both cost-effective and dependable. Montreal began construction in 1962 of its Indoor City, an interconnected network of pedestrian walkways, retail centers, residential areas, and public transportation—about half of which is underground. As of 2006, the structure extended almost 20 miles in length and covered an area of more than 4.5 square miles in Montreal's downtown core. The project has led to better access downtown, decreased walking distances, and made available additional available public space aboveground (El-Geneidy et al., 2011).

Urbanization is viewed by some as a primary cause of many of today's societal problems, but it is also viewed as a means to sustainably provide for the populations projected for the 21st century, according to participants in a recent National Research Council (NRC) workshop on urban sustainability research (Shaffer and Vollmer, 2010). While urbanization may not be a root cause, certain problems may have been compounded by it. Participants of that workshop identified a variety of factors that intensify the impacts of urbanization (prodigious consumption of resources in concentrated areas, environmental decline, public health problems, and economic and social inequalities) and reflect the failure of society to recognize urban areas as systems.

Shifting our image of a city from a dense set of autonomous people, structures, and infrastructure facilities to a dynamic system of interdependent elements is not a simple feat, but is essential to our capacity for resilience and ability to adapt to future challenges. An integrated three-dimensional approach to infrastructure design and management that considers and values space usage and human and social needs over time benefits all sectors of the community by protecting public health, reducing risks, maximizing reliability and long-term performance of urban infrastructure systems, and minimizing long-term costs.

The underground is a valuable resource. Urban planning too rarely takes a systematic account of the space both above and beneath Earth's surface on a coordinated basis at any large scale, and rarely incorporates infrastructure lifecycle planning or long-term infrastructure sustainability when deciding a future course. Under the sponsorship of the National Science Foundation, the NRC convened a new panel of experts to explore sustainable underground development in the urban environment, to identify research needed to make good use of the advantages, and to develop an enhanced public and technical community understanding of the role of engineering of underground space in the sustainability of the urban built environment. The committee comprised researchers and practitioners with expertise in geotechnical engineering, underground construction, trenchless technologies, risk assessment, and visualization techniques for geotechnical applications. Additionally, the committee included expertise in sustainable infrastructure development, infrastructure policy and planning, and fire prevention, safety, and ventilation in the underground. The committee's statement of task is provided in Box 1.1. Committee member biographies are included as Appendix A, and agendas from the committee's open session meetings are included in Appendix B.

DEFINING UNDERGROUND INFRASTRUCTURE

In general terms, urban infrastructure refers to all those physical and organizational structures that allow an urban system to function. Many types of infrastructure form the physical setting of the urban system (e.g., roads, utilities, buildings) and the governing, economic, and social frameworks that define a society. Underground infrastructure refers to any physical infrastructure that is placed beneath the surface and includes underground utilities (e.g., water, power, gas, communications, waste management), transportation (e.g., roads and highways, subways, freight and passenger rail) and their supporting facilities, building foundations, and any structure built in the underground to accommodate residential, industrial, manufacturing, recreational, or other purpose. Many types of infrastructure are further defined in Chapter 3. Given the broad nature of the committee charge and the many types of underground infrastructure, this report often generalizes underground infrastructure as a single category in many discussions, especially when referring to systems of infrastructure. It should be noted, however, that the benefits and challenges of individual types of underground infrastructure are not shared by all. Underground infrastructure is owned and operating by many different types of entities that serve many types

BOX 1.1
Statement of Task

An ad hoc committee of the National Academies will conduct a study to explore the potential advantages of underground development in the urban environment, to identify the research needed to take advantage of these opportunities, and to develop an enhanced public and technical community understanding of the role of engineering of underground space in the sustainability of the urban built environment, specifically the minimization of consumption of nonrenewable energy resources, construction materials, and negative impact on the natural, built, and social environments. In particular the study will:

- Summarize current geological and geotechnical engineering knowledge about underground development in the urban environment and how utilization of underground could increase sustainability, including knowledge of geologic site characterization, construction and geotechnical monitoring techniques, energy requirements, use of excavated materials, and lifecycle costs and benefits of underground infrastructure development.
- Identify the research needed to capitalize on opportunities for enhancing sustainable urban development through underground engineering, in the following areas:
    - Underground characterization, prediction of the geologic environment, and ground response critical for successful design and construction of underground projects and critical facilities to maximize sustainability and resiliency;
    - Construction and monitoring methodologies and enhanced excava-

of stakeholders, each with potentially different and sometimes opposing needs, interests, governing structures, and resources.

SUSTAINABILITY

Refining the definition of sustainability as it applies to underground development was the first task undertaken by the study committee. Earlier work illustrates the difficulty defining terms such as "sustainability" and even "urban" (e.g., Shaffer and Vollmer, 2010). The concept of "Sustainable Development" was described by the World Commission on Environment and Development in 1987 as "meeting the needs of the present without compromising the ability of future generations to meet their own needs" (UN, 1987). Terms such as "resilience" are often related to sustainability (e.g., NRC, 2011). The present study committee considers the maintenance of quality of life as part of sustainability, and it recognizes that incorporating sustainability into societal management practice must occur at many scales—from the global and national down to the individual project scale. Defining sustainability as part of implementable urban systems at the local level becomes more difficult because the term becomes infused with

tion methods, including tunneling, conducive to sustainable and resilient underground development;

- Smart underground structures and conduits that report their status;
- Health and safety considerations, such as cost-effective ventilation, light, and concerns related to radon exposure or fire control;
- Lifecycle cost and benefit issues, including reduced energy needs for heating and cooling, reduced construction material use, use of excavated materials, increased longevity of underground structures and reduced maintenance associated with stable temperatures and isolation from surface weathering effects;
- The potential sustainability benefits of increased use of underground space for human transportation systems, including roadways and mass transit, and freight;
- The potential for integrating of energy, water, and waste systems for certain urban regions to improve sustainability; and
- How underground development might address concerns related to the impacts of climate change on the urban environment.

The committee will recommend directions for a new underground engineering research track focused on earth systems engineering and management to ensure future human resources for sustainable underground development, will analyze the advantages and disadvantages of establishing a new research center in this area, and consider other potential options for enhancing the human resource capacity for sustainable underground development (including the status quo). The committee also will consider from a social science point of view, the policy, economic, and human behavioral drivers that promote or inhibit the development of the subsurface in a sustainable manner, but will not make policy or funding recommendations.

local values. The committee's definition of sustainable urban underground development is provided in Box 1.2.

The committee recognizes resilience as a key attribute of sustainability and defines resilience as the ability to respond to change in the environment—especially as a result of natural or human-caused disaster—with minimum impact to function. This is fairly consistent with definitions of resilience that appear in the social science literature (e.g., Norris et al., 2008). The ability to sustain expected societal services is a demonstration of resilience. In a societal setting, especially in the context of engineered systems, resilience is often associated with redundancy and reserves. However, the committee recognizes that resilience is more than the design of back-up systems and physical stockpiles. It encompasses a mindset in which society is considered a system where the underground plays a critical but often overlooked role.

In urban societies, the underground is part of a complex system that includes surface and above ground (e.g., bridges, skyscrapers) real estate. Without proper consideration of three-dimensional space and space usage over time, conflicts caused by competing use of the underground, or the problems associated with pollution of underground resources (e.g., space, groundwater, and materials) can

BOX 1.2

Definition of Sustainable Urban Underground Development

For the purpose of this report, sustainable urban underground development is an approach to subsurface development that meets current human needs while conserving resources and the natural and built environments to meet the needs of future generations. Sustainable urban underground development requires a systems perspective for above- and belowground resource use and management. Characteristics of sustainability as used in this report include consideration of cost effectiveness; longevity; functionality; safety; aesthetics and quality of life; upgradeability and adaptability; and the simultaneous maximizing of environmental and social benefits, resilience, and reliability, while minimizing potential negative impacts.

result. The resources of the urban underground need to be considered holistically for the most sustainable solutions (e.g., Parriaux et al., 2006). Individual projects are often framed independent of other planning and placed in the context of existing space use, rather than as part of long-term planning that allows integrated use of underground and surface space resources. Underground space is often not coherently or explicitly valued. As a result, most project designs are not chosen to preserve the opportunity for future flexibility and alternative uses or access. We have poor knowledge of the direct, indirect, and social costs of underground usage, and we have few metrics of the lifecycle benefits of investment in the underground.

Long-term sustainability is rarely a consideration in the early stages of the development of populated areas. An urbanization pattern observed in river valley settlements of developing countries serves as example of how human settlements can grow based on short-term and individual needs. For example, a hypothetical small settlement in a river valley may have plenty of room for both living and farming close to the river—typically the main water source. As the village grows, the fertile valley floor becomes significantly built over, and the adjacent hillsides—typically with poorer soil and requiring greater farming effort—are terraced for farming. Benefits of being close to the river are lost, and more difficult farming conditions are created. Quite different growth patterns may have evolved if long-term sustainability was considered from the outset.

A sustainability analysis might look at whether it would be better to terrace the hillsides for housing, providing greater flood protection in residential areas, and reserving the river valley for agriculture. Inherent in such an analysis would be consideration of which difficulties of outgrowing available land can be more easily solved—is it easier to create new productive agricultural land or to develop water supply and transportation approaches to service hillside developments? In real scenarios, such decisions extend to a regional and national context, but the example illustrates that human settlement systems do not necessarily evolve in their own best long-term interest.

HAZARD AND RISK

The terms *hazard* and *risk* appear throughout this report. There are many definitions of these terms, and even within the literature of a single discipline, the terms may be used inconsistently and interchangeably. Box 1.3 provides definitions for these terms as they are used throughout this report.

A BRIEF HISTORY OF UNDERGROUND OCCUPATION

To establish a perspective for present and future underground use, it is useful to summarize the centuries of past underground use. A rich legacy of fossil records and ancient tools, art, and structural ruins suggests that humans have had a complex and intimate association with the subsurface ever since evolving into modern *Homo sapiens*. Humans have sought practical shelter underground, but the underground seems to have evoked a sense of the supernatural and a desire for aesthetic expression (see Box 1.4). Human remains, shells, animal bones, and stone artifacts discovered in the Klasies River Mouth Cave in South Africa offer strong evidence that modern humans lived there more than 120,000 years ago when the climate was as warm or warmer than today (Rightmire and Deacon, 1991).

At the most basic level, the underground provided rock shelters and caves as refuge from harsh climates and mortal enemies, water and mineral reserves,

BOX 1.3
Definitions Associated with Hazard and Risk

The committee defines hazard as the potential to cause harm. These are threats to people, infrastructure, the environment, or social systems.

Sustainability is dependent on accounting for all sources of risk and all potential consequences, including some with impacts that are difficult to quantify. These may include social, environmental, and other less tangible long-term impacts that traditional engineering practice may not consider. The committee adopts the National Infrastructure Protection Program expanded definition of risk that include

> *the expected magnitude of loss (e.g., deaths, injuries, economic damage, loss of public confidence, or government capability) due to a terrorist attack, natural disaster, or other incident, along with the likelihood of such an event occurring and causing that loss (DHS, 2006).*

The committee defines vulnerability as the extent to which individuals, infrastructure, institutions, or systems can be harmed or damaged in the event of a hazardous event.

BOX 1.4
Underground Spirituality and Artistic Expression

There is an enduring influence of the underground on our collective imagination. The underground's wide-ranging literary and real-life associations with death and the afterlife, hidden demons and monsters, sacred rituals, heroic sagas, clandestine political rebellions, organized crime, anarchic music and theatre, film noir, adventure-seeking spelunkers, and the eternal search for precious metals and minerals reflect its power and paradoxical imagery. The underground has never been a neutral realm in terms of human perceptions and emotions.

Beyond basic survival, humans have been attracted to the underground over tens of thousands of years for spiritual and artistic expression, recreation, and religious ceremonies, especially in the commemoration of the dead. The evocative paintings and engravings of animals and hunting scenes set deep in the Chauvet-Pont-D'Arc Cave in southern France (see Figure) have been carbon-dated to more than 30,000 years ago. Vestiges of ancient underground temples, crypts, and ceremonial sites can be found throughout the world, including Chavin de Huantar in Peru, the Osireon (Strabo's Well) in Egypt, and the Hypogeum in Malta. Similarly, the mythologies of many cultures included gods and goddesses specifically dedicated to the underworld. The Roman version, Pluto, performed double-duty as the god of wealth because he also presided over all the precious metals hidden in the earth.

FIGURE Reproduction of a fresco found deep in the cave of *Chauvet-Pont-D'Arc* in southern France, drawn 30,000 years before present. SOURCE: The Cave of *Chauvet-Point-D' Arc*, available at http://commons.wikimedia.org/wiki/File:Paintings_from_the_Chauvet_cave_(museum_replica).jpg.

and ambient places to store food—all key factors for survival then as now. Some cultures have made the underground an integral part of daily life and their principal dwellings for thousands of years. Indigenous communities in China, Turkey, Spain, and Tunisia have continuously occupied man-made spaces belowground for more than 4,000 years; tens of millions of present-day Chinese still live in

FIGURE 1.1 Example of a multistory *yao dong*, a type of cave dwelling carved into vertical or near vertical walls of loess (a silty soil), in the Shaanxi province in northwestern China. Approximately 90 percent of rural dwellers in the region lives in *yao dong*. SOURCE: Liu, 2009. License CC BY-NC-SA 3.0.

dwellings known as *yao dong* (see Figure 1.1) carved into vertical walls of loess (a silty soil), many of which are said to date back to 5000 B.C. (Golany, 1996; Meijenfeldt, 2003).

Engineers of the ancient world skillfully exploited the underground with rudimentary technology to promote the growth of emerging cities and commerce. The first water supply technology in Jerusalem was an underground water system constructed during the Middle Bronze Age (2000-1500 B.C.) for both domestic and agricultural purposes (Barghouth and Al-Sa'ed, 2009). The 1,036 meter Tunnel of Eupalinos, the first-known deep tunnel in history, was part of the water supply system of the island of Samos in Greece and named after the engineer who designed and constructed it in 530 B.C.; it operated for nearly 1,000 years until the fifth century A.D. (Koutsoyiannis et al., 2008). The spectacular Roman cistern, Piscina Mirabilis (Figure 1.2), with a volumetric capacity of 12,000 cubic meters of water, was carved out of a tufa (a soft porous volcanic rock) hill in the Campania region in Southern Italy during the reign of Emperor Augustus Caesar between 33 and 12 B.C.E. to provide fresh water for an important Roman naval base as well as several major cities and ports (De Feo, 2008).

Much of the world's population relies on the underground as a matter of daily necessity, convenience, or aesthetic choice. A small percentage lives or works underground full-time; a significantly larger share occasionally occupies the underground to attend concerts or movies, shop, worship, park vehicles, store things, or find relief from severe surface weather conditions. A frequent means of direct human contact with the underground is travel through it via automobile or railway tunnels, transit tubes, or pedestrian passageways. Many contempo-

FIGURE 1.2 The *Piscina Mirabilis* in southern Italy was a 12,000 cubic meter capacity cistern carved by the ancient Romans between 33 and 12 B.C.E. SOURCE: Ra Boe/Wikipedia, License CC by-sa 3.0, available at http://en.wikipedia.org/wiki/File:Piscina_Mirabilis_2010-by-RaBoe-18.jpg.

rary underground facilities are world-renowned cultural icons, including the Moscow Metro (Figure 1.3), the *Carrousel du Louvre* in Paris (Figure 1.4), the Glass Temple in Kyoto, Japan, Philharmonic Hall in Cologne, Germany, and the *Cathedral Metropolitana* in Brasilia, Brazil.

Much of the history of underground construction is contemporary with the history of tunneling. For general accounts of the history of underground engineering, the reader is referred to work by Sandström (1963), Széchy (1970), Harding (1981), and Wood (2000).

FIGURE 1.3 Underground Metro platform in Moscow. SOURCE: Boris Kogut. Reprinted with permission of Boris Kogut ©2012.

FIGURE 1.4 The inverted pyramid in the *Carrousel du Louvre*, an underground shopping mall in Paris, France, adjacent to the Louvre museum of fine art. The underground facility accommodates shopping, live theatre, auditorium space, parking, and underground access to the famous museum. The inverted pyramid is made of glass and allows natural light into the underground facility. SOURCE: Photo by Gard Karlsen, available at http://gardkarlsen.com.

TABLE 1.1 Examples of Potential Benefits and Drawbacks of Underground Space

Major Issues	Sub Category	Potential Benefits	Potential Drawbacks
Physical and Institutional Issues	Location	Proximity for functional benefit Limited use of surface space Provides utility and transportation services	Unfavorable geology in chosen location Uncertain geology
	Isolation	Climatic: thermal, severe weather, fire, earthquake Protection: noise, vibration, explosion, fallout, industrial accident Security: limited access, protected surfaces Containment: hazardous materials and processes	Climatic: thermal, flooding, Communication Human issues: pyschological concerns, fire safety, personal safety
	Preservation	Aesthetics: visual impact, interior design Environmental: natural landscape, ecology Low material degradation	Aesthetics: visual impact, building services, skillful design required Environmental: site degradation, drainage, pollution
	Layout	Topographical freedom 3-dimensional planning	Ground support Span limitations Access limitations Adaptability Sewage removal
	Institutional		Easement acquisitions Permits Building code Investment uncertainty
Life-cycle cost	Initial Cost	Land cost savings Construction savings: no structural support, weather independent, scale of construction Sale of excavated materials or minerals Savings in specialized design features	Confined work conditions Ground support Limited Access Ground excavation, transport and disposal Cost uncertainty: geological, contractual, institutional delays
	Operating Cost	Maintenance Insurance Energy Use	Equipment/materials access Personnel access Ventilation and lighting Maintenance and repair

Major Issues	Sub Category	Potential Benefits	Potential Drawbacks
Societal Issues		Land use efficiency Transportation and circulation efficiency Energy conservation Environmental/aesthetics Disaster readiness National security Less construction disruption	Environmental degradation Irreversibility High embodied energy

SOURCE: Adapted from Carmody and Sterling, 1993.

POTENTIAL BENEFITS AND CHALLENGES ASSOCIATED WITH DEVELOPING UNDERGROUND SPACE

Underground space development presents many potential benefits, but there are many challenges to overcome in designing, operating, and maintaining underground infrastructure so that it contributes to urban sustainability. Table 1.1 lists some of the potential benefits and disadvantages of underground space development. Urban development patterns set in motion are hard to change. Underground space is often engineered to meet the needs of a single project or use. Design sometimes doesn't accommodate long-term maintenance, much less interactions with existing or future structures. Many past and current utility layout practices, for example, are not consistent with sustainability goals (see Box 1.5) and do not take into account long-term impacts on the environment, economy, society, natural resources, or governance. As described by Sterling et al. (2012), underground facilities can influence the ways in which human occupancy of a land affects the surface environment as well as the economic and social structures of an urban area in ways not possible using already existing surface structures. Properly planned and maintained, underground infrastructure can contribute to sustainability by preserving natural surface resources (e.g., land, water, biodiversity), reducing air pollution related to transportation, creating opportunities for less energy use and waste generation, and creating structures more resilient to many catastrophic events. Examples worldwide demonstrate how underground facilities can have low environmental impact. The Groene Hart Tunnel that lies underground between the four largest cities in the Netherlands, for example, has provided rapid connection from Amsterdam to major economic centers in Europe without detriment to the large green space of Groene Hart (Sabel Communicatie, 2007; ITA-AITES, 2011).

The decision to move societal features underground is a major step in the development of human settlements. Infrastructure is often placed underground if it cannot fit or is not wanted at or above the surface. The decision to build underground may be made, for example, when contemplating a new transit system in a historic city with a unique and culturally important surface environ-

BOX 1.5

Sustainability of Underground Utility Design

Long-term sustainability of infrastructure design, such as for essential urban utilities, has rarely been considered in the past and is only sometimes considered today. Figure 1 shows what may well have been an engineering design feat in 1917. A "spaghetti" of underground pipes and conduits provided for a variety of services; however, repair or replacement of any element of this infrastructure would likely have resulted in disruption to local traffic and infrastructure service, and possibly in damage to other elements of the infrastructure. Utility corridors called utilidors, on the other hand, are enclosed conduits employed by some urban areas designed to carry multiple utility lines such as electrical, water and sewer, and communications (see Figure 2). Repair of individual utility lines can be conducted with minimal interference to surface structures or other infrastructure. Design can accommodate multiple levels of utilidors (see Figure 3). Further discussion on utilidors, their benefits, and barriers to their use is provided in Chapter 3.

FIGURE 1 The placement of underground utility infrastructure on Wall Street (circa 1917). SOURCE: Consolidated Edison Company of New York, Inc. Reprinted with permission from Con Edison Company of New York.

ment, or where existing street layouts or traffic levels do not permit new surface or elevated alignments. However, a desired location may present challenges—structures may already exist in the underground space, or geologic conditions may not be ideal. Urban needs often trump favorable geology. Although there is a large volume beneath Earth's surface, perhaps only the first 30 meters beneath cities are used to support most urban functions. And of the first 30 meters, the vast majority of subsurface utilities and transportation services are placed beneath

FIGURE 2 Example of a utilidor in Amsterdam that can carry multiple utility lines such as electrical, water and sewer, and communications. SOURCE: Courtesy H. Admiraal.

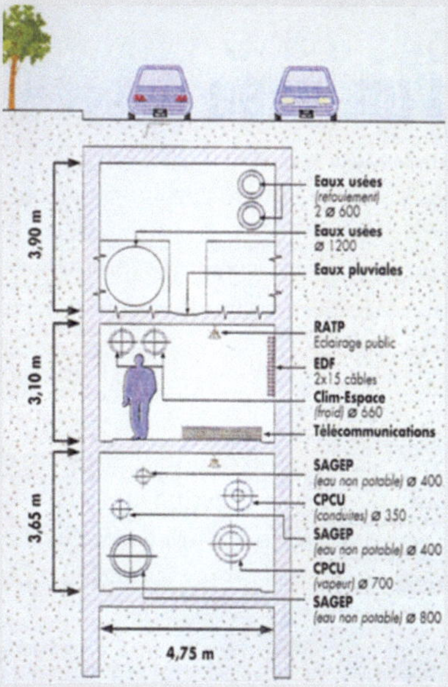

FIGURE 3 Schematic showing utilidor design in Paris, France. Multiple levels of utilidor can be accommodated. SOURCE: SEMAPA. Reprinted with permission from © SEMAPA.

public rights-of-way (e.g., streets and sidewalks). Additionally, once disturbed, the underground cannot be restored to its prior condition. This is particularly true for spaces such as bored tunnels or caverns created within soil or rock; their presence significantly affects future options and costs of new underground infrastructure in their vicinity.

Structural and geotechnical constraints can limit the types of facilities placed underground in a given location or increase construction or operational costs relative to cost for surface facilities. Water and moisture control in underground space is challenging—underground infrastructure needs to be protected from inflow or seepage of unwanted fluids, and vulnerable groundwater resources need to be protected from contamination and depletion. Existing underground infrastructure or legacy construction debris constrain underground planning and construction. However, placing infrastructure underground provides an added development dimension: complex transportation systems can be located beneath cities, and tunnels can be placed beneath mountain ranges and rivers.

HUMAN FACTORS AFFECTING UNDERGROUND DEVELOPMENT

Another set of opportunities and challenges are those associated with people using or working in underground space. These include institutional and administrative constraints related to planning and permitting, underground infrastructure security, safety, and the psychological acceptability of underground structures and their use. This report does not explore all these issues in great detail, but Chapter 4 provides more discussion of these issues. Simply not having an experiential basis for decision making related to underground infrastructure makes these issues more challenging. Underground permitting, for example, is less routine than for surface facilities and therefore can be more cumbersome. Safety codes for occupied underground facilities, including codes related to fire, egress, and ventilation systems, may not exist or may be inadequate (see Chapter 4 for discussion on existing codes for certain facility types). Underground infrastructure can be more secure than surface infrastructure because of the controlled access and isolation the underground offers. Similarly, the underground can be used to separate or isolate hazardous materials such as raw sewage or high-voltage electrical lines from people and infrastructure on the surface. On the other hand, that same separation means that protecting against physical hazards such as flooding, internal fire, and explosions is more challenging, especially as diverse underground infrastructure becomes more integrated with other underground and surface infrastructure.

Access to underground facilities or resources may be difficult or impossible for physically impaired individuals without mechanical conveyance. Safety for people with special needs is a major challenge, for example, in the event of power failure. Other members of society may simply be uncomfortable with the notion of the underground, or they may find the lack of natural light in the underground unpleasant or spatially disorienting. And for some, there are physiological or psychological barriers to working, living, wayfinding and commuting, or playing underground including claustrophobia or fear of isolation. Many with discomforts may learn to use and appreciate the underground with appropriate public education campaigns. Discomforts can be effectively addressed with skillful planning, innovative designs, layout, finish, and lighting.

Cautionary tales of underground communities created by a drive for efficiency or a response to a calamity can be found in a number of literary works (e.g., Forster, 1909). Such concerns need to be considered—both in broad terms of what living and working environments should be—as well as in the details of facility design.

The balance between the desire for open air living and the convenience or protection offered by underground facilities is not a fixed point. Although a small percentage of the population may be unable psychologically to tolerate underground facilities, others choose cave exploration as a hobby. Most in society, perhaps, are influenced by a conscious or unconscious evaluation of the benefits and drawbacks relating to particular circumstances, for example, a fast, convenient journey on an underground metro versus a slow journey in a car or bus on the street, or shelter during a wartime attack. Good design in response to an understanding of what makes underground spaces interesting, attractive, safe, cost effective, and part of sustainable development within existing physical limitations can shift the balance point regarding perception of underground use.

REPORT ORGANIZATION

Daily urban life generally proceeds without residents noticing the operation of underground infrastructure, and perhaps the success of infrastructure may be measured, in part, by how much it is taken for granted. Engineers design and build for function while minimizing risk. However, it is impossible to completely eliminate risk. Failures of infrastructure will happen as a result of age, error, or extreme events. It is such failures that lead to the need for reports such as this, which describes many types of infrastructure failures to illustrate the challenges to be overcome. Underground infrastructure successes are also highlighted to demonstrate approaches to underground engineering that may contribute to sustainable urban development.

Countries such as Finland, Sweden, Norway, the Netherlands, Japan, China, and Singapore have taken national-level action that promotes underground space use as a policy issue. Countries such as France, the United Kingdom, the United States, and Germany have significant levels of underground activity, but underground use lacks a national level of attention (Sterling et al., 2012). In this report, the committee will argue that a multilevel, multidisciplinary approach to urban planning that incorporates underground engineering as part of the overall approach may provide a better framework for sustainable urban development.

The statement of task as it appears in Box 1.1 is long and broad, but after considerable study of the task, and following multiple discussions with the committee sponsor, the committee came to understand that the heart of its task is consistent with the committee's given title: the Committee on Underground Engineering for Sustainable Development. The committee deliberated its charge and prepared this report considering the contributions of engineered underground space to sustainable development as well as what is needed in the social, educational, regulatory, educational, and research environments to allow those contributions to be made.

This report is organized into seven chapters. Chapter 2 traces the evolution of urban underground space use and the drivers affecting proper development. In Chapter 3, the committee discusses the role of underground engineering in sustainability and some of the challenges of sustainable underground development. Chapter 4 examines human-technical system relationships and the hazards related to human use of underground space. The assessment of costs and benefits of underground infrastructure and lifecycle sustainability are addressed in Chapter 5. Chapter 6 explores the technologies that make underground engineering possible and discusses the types of innovations that could increase the contributions of underground engineering to sustainable development. Finally, the committee presents its overarching conclusions in Chapter 7 in the context of a framework to improve institutional, educational, research, and workforce capacities for underground engineering for sustainability.

REFERENCES

Barghouth, J.M., and R.M.Y. Al-Sa'ed. 2009. Sustainability of ancient water supply facilities in Jerusalem. Sustainability 1(4):1106-1119.

Carmody, J., and R.L. Sterling. 1993. Underground Space Design: A Guide to Subsurface Utilization and Design for People in Underground Spaces. New York: Van Nostrand Reinhold.

De Feo, G. 2008. The Piscina Mirabilis in Miseno, Southern Italy. Pp. 11-14 in Specialist Group on Water and Wastewater Technologies in Ancient Civilizations Newsletter, May 2008. International Water Association [online]. Available: http://www.miraproject.eu/MELIA/copy_of_about-melia/IWA Ancient Civilisations Newsletter.pdf (accessed March 31, 2011).

DHS (U.S. Department of Homeland Security). 2006. National Infrastructure Protection Plan: Building a Safer, More Secure, and More Resilient America. U.S. Department of Homeland Security [online]. Available: www.dhs.gov/xlibrary/assets/NIPP_Overview.pdf (accessed March 30, 2011).

El-Geneidy, A., L. Kastelberger, and H. Abdelhamid. 2011. Montreal's roots: Exploring the growth of Montreal's indoor city. Journal of Transport and Land Use. 4(2):33-46.

Forster, E.M. 1909. The Machine Stops [online]. Available: http://archive.ncsa.illinois.edu/prajlich/forster.html (accessed April 30, 2012).

Golany, G.S. 1996. International uses of geo-space: Indigenous and contemporary examples. Pp. 311-337 in Geo-Space Urban Design. New York: John Wiley & Sons.

Harding, H. 1981. Tunneling History and My Own Involvement. Toronto, Canada: Golder Associates.

IPCC (Intergovernmental Panel on Climate Change). 2007. Climate Change 2007: Synthesis Report. Geneva, Switzerland: Intergovernmental Panel on Climate Change [online]. Available: www.ipcc.ch/publications_and_data/publications_ipcc_fourth_assessment_report_synthesis_report.htm (accessed October 14, 2010).

IPCC. 2008. Climate Change and Water: IPCC Technical Paper VI. Geneva, Switzerland: Intergovernmental Panel on Climate Change [online]. Available: http://www.ipcc.ch/pdf/technical-papers/climate-change-water-en.pdf (accessed March 30, 2011).

Isgård, E. 1975. Underground Wastewater Treatment Plants. Journal of the Water Pollution Control Federation. 47(4):684-687.

ITA-AITES (Association Internationale des Tunnels-International Tunnelling and Underground Space Association). 2011. Protection of Wildlife [online]. Available: http://www.ita-aites.org/index.php?id=120 (accessed April 30, 2012).

Kammen, D.M., and A. Jacobson. 2006. Science and Engineering Research That Values the Planet. Pp. 59-70 in Frontiers of Engineering: Reports on Leading-Edge Engineering from the 2005 Symposium. Washington, DC: The National Academies Press.

Koutsoyiannis, D., N. Zarkadoulas, A.N. Angelakis, and G. Tchobanoglous. 2008. Urban water management in ancient Greece: Legacies and lessons. Journal of Water Resource Planning and Management 134(1):45-54.

Linger, D.A., G.H. Baker, and R.G. Little. 2002. Applications of underground structures for the physical protection of critical infrastructure. Pp. 333-342 in North American Tunneling 2002, L. Ozdemir, ed. Lisse, The Netherlands: Swets & Zeitlinger.

Liu, J. 2009. Yaodong - Cave Dwellings on Loess Plateau. ChinaBlog.cc. [online]. Available: http://chinablog.cc/2009/02/yaodong-cave-dwellings-on-loess-plateau/ (accessed March 31, 2011).

Markham, V.D. 2008. U.S. Population, Energy & Climate Change. Center for Environment & Population [online]. Available: http://www.cepnet.org./documents/US-Population-Energy-Climate_Change2009.pdf (accessed October 14, 2010).

Massachusetts Turnpike Authority. 2006. Economic Impact of the Massachusetts Turnpike Authority & Related Projects, Volume I: The Turnpike Authority as a Transportation Provider. Massachusetts Department of Transportation [online]. Available: http://www.massdot.state.ma.us/portals/0/docs/infoCenter/additionalDocs/MTA-Economic-V1.pdf (accessed April 27, 2012).

Meijenfeldt, E.V. 2003. Below Ground Level: Creating New Spaces for Contemporary Architecture, M. Geluk, ed. Basel, Boston: Birkhauser.

NAE (National Academy of Engineering). 2000. Greatest Engineering Achievements of the 20th Century [online]. Available: http://www.nationalacademies.org/greatachievements/List.PDF (accessed July 9, 2012).

Norris, F.H., S.P. Stevens, B. Pfefferbaum, K.F. Wyche, and R.L. Pfefferbaum. 2008. Community resilience as a metaphor: Theory set of capacities, and strategy for disaster readiness. American Journal of Community Psychology 41(1-2):127-150.

NRC (National Research Council). 2010. Adapting to the Impacts of Climate Change. Washington DC: The National Academies Press.

NRC. 2011. Building Community Disaster Resilience through Private-Public Collaboration. Washington DC: The National Academies Press.

Papay, L.T. 2002. Urban design: The grand challenge. Pp. 91-94 in Engineering and Environmental Challenges: Technical Symposium on Earth Systems Engineering. Washington, DC: National Academy Press.

Parriaux, A., L. Tacher, V. Kaufmann, and P. Blunier. 2006. Underground resources and sustainable development in urban areas. Paper number 700 in Engineering Geology for Tomorrow City: International Association for Engineering and Geology 2006: The 10th International Congress, September 6-10, 2006, Nottingham, U.K. [online]. Available: http://www.iaeg.info/iaeg2006/start.htm (accessed March 30, 2011).

Rightmire, G.P., and H.J. Deacon. 1991. Comparative studies of Late Pleistocene human remains from Klasies River Mouth, South Africa. Journal Human Evolution 20(2):131-156.

Sabel Communicatie. 2007. Guide to Good Practice: Underground Space. Rotterdam, The Netherlands: Europoint Media [online]. Available: http://www.scribd.com/doc/9663895/Connected-Cities-guide-to-good-practice-underground-space (accessed April 30, 2012).

Sandström, G. 1963. Tunnels. New York: Holt, Rinehart and Winston.

Shaffer, D., and D. Vollmer. 2010. Pathways to Urban Sustainability: Research and Development in Urban Systems. Washington, DC: The National Academies Press.

Sterling, R.L., and J.P. Godard. 2000. Geoengineering considerations in the optimum use of underground space. Pp. 708-720 in GeoEng2000: An International Conference on Geotechnical & Geological Engineering, Vol.1: Invited Papers. International Tunnelling and Underground Space Association [online]. Available: http://www.itaaites.org/fileadmin/filemounts/general/pdf/ItaAssociation/ProductAndPublication/ItaPositionPapers/Geoingineering/geo202.pdf (accessed April 6, 2011).

Sterling, R., H. Admiraal, N. Bobylev, H. Parker, J.P. Godard, I. Vähäaho, C.D.F. Rogers, X. Shi, and T. Hanamura. 2012. Sustainability issues for underground spaces in urban areas. Proceedings of ICE - Urban Design and Planning. Available at www.icevirtuallibrary.com/content/article/10.1680/udap.10.00020.

Széchy, K. 1970. The Art of Tunnelling. Budapest: Akademia Kiado.

UN (United Nations). 1987. Report of the World Commission on Environment and Development. General Assembly. A/RES/42/187. United Nations [online]. Available: http://www.un.org/documents/ga/res/42/ares42-187.htm (accessed April 6, 2011).

U.S. Census Bureau. 2011. Thc 2012 Statistical Abstract: The National Data Book [online]. Available: http://www.census.gov/compendia/statab/cats/population.html (accessed August 4, 2011).

Wood, A.M. 2000. Tunneling: Management by Design. London: E & FN Spon.

2

The Evolution of and Factors Affecting Underground Development

Underground space use has evolved as villages and towns have grown into cities. Water, energy, sewage, and wastes that were once carted in and out of town on street surfaces are now transported via underground conduits. Urban dwellers are often unaware of the connections between essential utility services and structures in the buildings where they live and work and the underground supporting infrastructure and services. Consequently, there is a lack of public appreciation of how critical underground resources are to the proper functioning and high standards of living in U.S. urban areas.

The underground has always provided physical foundation support for buildings and other surface structures. Early building foundations may have been simple sets of stones selected and placed by hand into shallow excavations. Today, foundations for large buildings and skyscrapers may include deep pilings, conduits for geothermal heating and cooling, and multiple levels of basement space that may provide, for example, shopping concourses, underground parking, utility plants, and high-quality storage. Well-designed foundations take into account the soil, rock, groundwater, and other site-specific conditions, and help buildings resist major seismic and extreme wind effects. Hard-won experience, artful skills, and knowledge from many science and engineering disciplines contribute to the development of the processes and procedures used today to site, design, and build large structures.

Although cities grow upwards and outwards, their growth is dependent on underground building foundations and utility infrastructure. In most municipalities, planning and zoning of surface and air spaces are through local governments. Unfortunately, underground space is not similarly planned and zoned, and an explicit value for underground space is not generally recognized (Sterling et al.,

2012). Formal planning and control of underground space by municipalities is a responsibility to be recognized and acted upon in the United States if sustainable urban development is to be realized. In some countries such as China, planning of underground space is a special focus for responding to urban growth, and such plans have been developed by almost every large Chinese city over the past few years (Guo et al., in press).

This chapter traces the evolution of urban underground space and illustrates how the progressive and piecemeal development of underground space poses significantly more restrictions on future development than in the cases of surface facilities and infrastructure development.

EXPANSION OF THE UNDERGROUND IN THE PAST CENTURY

Sewage systems are placed underground to use gravity to drain sewage away from buildings. Water distribution systems are often placed underground to protect them against freezing and other damage. Telecommunications and electric power supply systems may be placed below ground according to local precedent, in consideration of the value placed on maintaining a secure and resilient infrastructure, for reasons related to surface aesthetics, or to minimize the effect of installation on property values. Concern regarding uncoordinated planning of underground space is not new. In 1914, George Webster, chief engineer and surveyor of Philadelphia, lamented that few large cities planned the space beneath streets, or charted the utilities and services placed there (Webster, 1914). He noted the importance of understanding what the underground was required to accommodate and discussed the need to plan for

- water, hot water, steam, sewer, refrigerating, and gas pipes; electrical conduits; pneumatic tubes; and as yet undetermined future services;
- galleries for pipes and conduits;
- vaults under sidewalks in the public right-of-way as a part of new building construction;
- subways for transit systems and passengers;
- tunnels beneath underground services to accommodate movement of people between business establishments without the need to cross streets or venture into weather; and
- underground freight movement services to connect freight terminals with commercial businesses and industrial establishments.

Webster advocated that underground space should be planned to facilitate future installations and minimize the costs and delays caused by future installations. He advocated for an official authoritative body to regulate underground usage, and he predicted that without such controls new large underground installations would come at greater cost and challenge when engineers were forced to work around existing infrastructure.

As we progress into the 21st century, the underground is used for all the purposes listed above, and many of the problems predicted a century ago have been realized. Table 2.1 describes the estimated lengths of major underground utility services in the United States, totaling approximately 10.8 million miles (17.4 million kilometers). Underground infrastructure has expanded to accommodate growing populations and new infrastructure services (and their multiple providers) but is still installed beneath the same public rights-of-way. As traffic becomes more congested with population growth, underground utility work that must be accessed from the surface results in increased traffic problems and expense. It has been reported that approximately 4 million holes are dug in the United Kingdom's roads and sidewalks by utilities at a cost of approximately \$2.25 billion[1] per year and consequent indirect costs of approximately \$4.5 billion per year (Farrimond, 2004). Analogous costs in the United States could well be many times larger.

Wastewater systems have also been expanded and the underground now accommodates large wastewater transport systems (e.g., sanitary and stormwater sewer systems; combined sewer systems) and combined sewer overflow (CSO) interceptor and storage tunnel systems with large diameter openings. Most segments of wastewater and drainage systems are designed to flow by gravity through pipes and tunnels and are therefore dependent on closely controlled vertical alignments. These systems are generally placed beneath the hodgepodge of existing shallow utility infrastructure, and they may block usage of that underground space for future services including rapid transit subways and high speed rail (HSR). Protecting access opportunities for such services argues for planning and permitting with a goal of preserving underground corridors for major high-value urban infrastructure. Foresight is vital to sustainability because such complex infrastructure is often not needed until much later in a city's evolution.

ENGINEERING THE UNDERGROUND FOR SUSTAINABILITY

Tunneling, a component of many underground construction projects, shares many properties with other types of construction done in urban societies. Certain challenges, however, may become amplified in an underground setting (Wood, 2000). For example

- there is greater dependence on the ground and understanding ground properties in terms of risk (see Box 1.3) to the construction project itself, other infrastructure, worker health and safety, the environment, and economic interests;
- there is higher interdependence between planning and project design

[1]Based on 2008 exchange rates.

TABLE 2.1 Estimated Lengths of Major U.S. Underground Utility Services

Transmission (miles)	Distribution/ Collection (miles)	Service (miles)	Total (miles)
Gas Gathering ....41,000 (DOE, 2006) Interstate ....250,000 Intrastate 75,000	1,212,688 (PHMSA, 2005)	780,392 (PHMSA, 2005)[a]	2,359,080
Hazardous Liquid.............160,868 (PHMSA, 2003)			160,868
Oil Gathering ....35,000 (Pipeline 101, 2001) Crude................................65,942 (BTS, 2004) Product 76,258			177,200
Water ....660,000 (Brongers, 2002)	995,644 (EPA, 2007)	854,364 (EPA 2007)[b]	2,510,008
Sewer Public Private	 724,000 (EPA, 2006) 500,000		1,224,000
Electric ....167,643 (NERC, 2006)	600,000[c]	400,000[d]	1,167,643
Telecom Underground Cable Metallic Fiber Buried Cable Metallic Fiber Conduit System Trench	 382,472 (FCC, 2006) 217,266 2,178,320 217,322 199,541		3,194,921
Grand Total			**10,793,719**

[a]The total number of gas services in the United States, according to PHMSA (2005), is 63,523,945. This number was then converted to miles by taking an average length of one service line to be 65 ft.
[b]The total number of water services in the United States, according to EPA (2007), is about 69,545,307. This number was then converted to miles by taking an average length of one service line to be 65 ft.
[c]Eleven U.S. utilities reported a total of 296,093 miles (Sterling et al., 2009). However, the length of underground electrical distribution is expected to be much less than for gas or water, which are fully underground. A figure of 600,000 miles is assumed as the U.S. total.
[d]This figure is a rough estimate based on underground electric service being less than half the length of underground water services.
SOURCE: Adapted from Sterling et al., 2009.

that arises from the need to stabilize the ground and exclude groundwater or contaminants;

- there are potentially fewer construction methods available given geologic and anthropogenic constraints;
- logistics can be more challenging because of restricted access and addressing worker safety (workers may be great distances from access points); and
- the expertise and time involved from project inception and completion can be great and may include that associated with community buy-in of a project and government compliance issues.

An underground project requires a systems perspective, such as illustrated in Figure 2.1, that emphasizes interactions between interrelated systems including those associated with land use, intermodal transportation, environmental, cultural, and socio-economic systems. This type of approach highlights the unique combination of skills, knowledge, management, and leadership required for successful infrastructure planning, construction, operation, and maintenance for a sustainable urban environment. Figure 2.1 represents a good start to the kind of thinking necessary, but sustainability of engineered systems within urban systems needs to be designed for much greater complexity and adaptability, such as is done for Complex Adaptive System of Systems (CASoS) engineering. CASoS

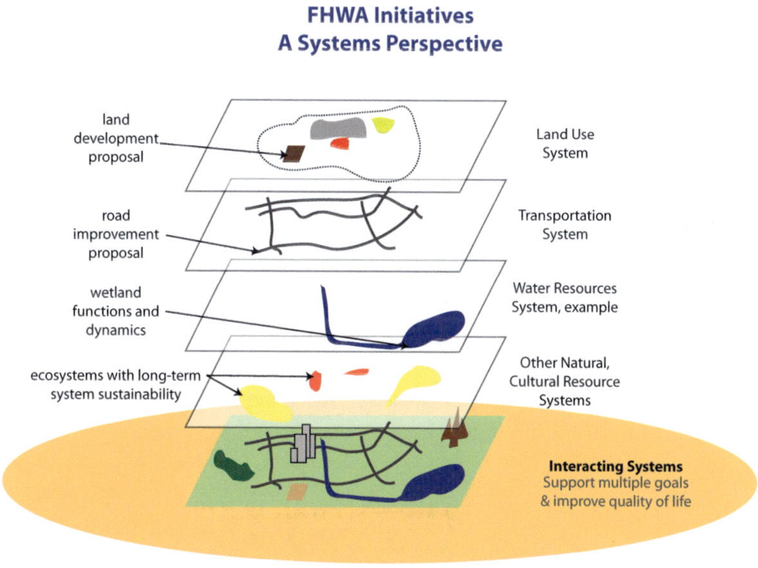

FIGURE 2.1 A systems perspective toward a foundation of interacting systems (shown at bottom/base of this graphic) that includes land use, intermodal transportation, natural, cultural, and socio-economic systems deliver quality of life and multiple benefits for the long-term. SOURCE: FHWA, 2008.

BOX 2.1

Complex Adaptive Systems of Systems Engineering

An initiative at Sandia National Laboratory has been the development of an engineering framework to solve large complex problems that combine physical, social, and technical systems called Complex Adaptive Systems of Systems (CASoS). CASoS broadly include physical infrastructure, government, people, and ecosystems. They are complex, real or abstract entities composed of systems, and change over time because of interactions within the system or environment (Glass et al., 2008). CASoS engineers use a set of defined iterative processes to solve problems, exploit opportunities, achieve goals, or answer questions in consideration of choices, intended and unintended costs and benefits, uncertainties, and how the system might be altered to yield better outcomes. Bringing about change in CASoS can be accomplished using conceptual models, system measurements, observational and experimental design, pattern recognition, policy investigation, engineering processes, real-time problem definitions (especially in times of crisis), and communication, and building the required intellectual capacity to conduct CASoS engineering focused on applications (Glass et al., 2008). The CASoS framework includes designing a computational model for the context, implementing the model in an actual environment, and reviewing actions at each step for correction, adaptation, and "fit performance" at each stage of action. The figure is a simplified diagram of the elements to be considered in CASoS engineering.

engineering considers the interdependencies and vulnerabilities of systems to reduce risk and maximize security and health (Glass et al., 2011), as described in Box 2.1.

Given such systems of systems approaches, the team that designs, constructs, and manages underground infrastructure needs to be interdisciplinary, and specific expertise will be required to respond to specific challenges (see Appendix C). However, it will be necessary for team members to be able to understand how each component of the project is part of a system of systems.

POLICY, ECONOMIC, AND HUMAN BEHAVIORAL DRIVERS THAT INFLUENCE DECISION MAKING

Electric power lines were already being buried in New York in the late 1800s (Schewe, 2007), but overhead electric lines are still common in cities across the United States. What factors drive acceptance of underground placement of infra-

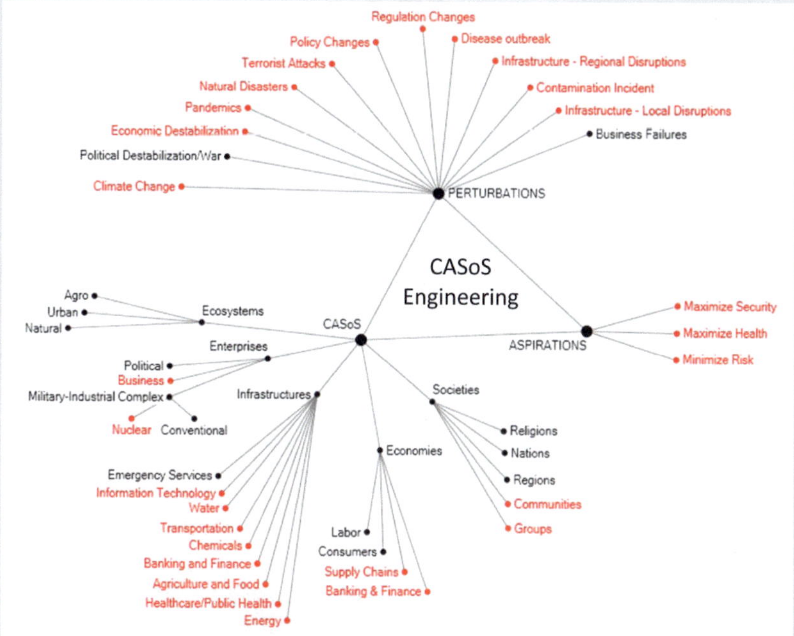

Figure Simplified diagram developed at Sandia National Laboratories of CASoS engineering application space as a simplified network. The diagram illustrates how CASoS engineering considers the relationships of the CASoS, the goals of engineering (termed aspirations), and elements that can influence the system (perturbations). Items in black represent existing applications for a specific CASoS and those in red represent those in development. SOURCE: Glass et al., 2011.

structure? In the past century, acceptance of underground utility installation has evolved to be based on a combination of environmental, cost, and performance issues. Long-term performance of underground facilities has yet to be quantified or demonstrated, yielding a source of uncertainty and unknown risk for decision makers. Triple-bottom-line cost estimates—analyses of social, environmental, and economic costs and benefits—for underground facilities may provide persuasive justification for underground installation, but direct and indirect impacts need to be considered for a true lifecycle engineering design.

Higher costs of underground utility installation may make the underground less attractive to the private sector, and government stakeholders often display mixed acceptance to underground installations, sometimes depending on their relationships with utility providers. The long-term outlook of community decision makers has a role in the acceptance of underground utilities. A decision to bury utilities is best made based on real costs and experience, rather than whether stakeholders "like" underground facilities. Technological advances in underground installation processes, system monitoring, and in the development of cost-effective utilidors (underground utility corridors that house multiple utilities such as water, sewer, power, and telecommunications) create advantages and incentives to place utilities underground.

Decision making related to the placement of major transportation systems often occurs among people with competing interests. Political views, community lifestyle preferences, and commercial interests in design and construction contracts influence decision making in many communities in the United States. Disruptions related to construction and operation and the impact of projects on taxes can dissuade public acceptance for underground installation. The public may also be concerned about security, fearing that mass transportation systems may allow access to a neighborhood for a large number of unknown people.

Government officials may be concerned with the "success" of a project—that major cost overruns and construction issues are limited and that the finished infrastructure is perceived by the public to have been a wise investment. Some politicians may be concerned that successful completion occurs before the next election cycle to reserve credit for success to the incumbents. Negative and positive experiences of other cities may influence how costs and risks of design options are accepted. Unfortunately, there are few detailed follow-up assessments of major infrastructure investments with data suitable for triple-bottom-line analyses. Without such information, too much focus may be placed on initial cost and too little on long-term performance and urban benefits.

The decision to place technical systems such as energy-related facilities, roads and railroads, shopping centers, waterworks, and wastewater treatment underground is based, often primarily, on technical data related to operational and environmental considerations, and considerations associated with safety, hygiene, disaster prevention, land use, and maintenance costs. Scandinavian experience with underground sewage treatment plants and hydropower facilities, for example, has led to a strong preference for underground infrastructure by the public, utility company, and government stakeholders, driven by the climatic, topographic, and geological environments (Parker, 2004).[2] U.S. efforts to develop underground facilities have been modest in comparison; adoption of new approaches is often inhibited by existing administrative controls, design guidelines, codes of practice, and labor practices (NRC, 2011).

A systematic analysis of the networks of decision makers and how the flow of information through the networks facilitates or inhibits decision making may be informative and a powerful tool if carefully applied (e.g., Butts, 2009). Whereas the number of stakeholders indicates that the web of networks in the case of urban system analysis is complex, even complex networks are not random

[2]For example, the Høvringen and Ladehammeren underground sewage treatment plants in Trodheim, Norway (Nordmark, 2002; Broch, 2006); the Skullerud water treatment plant in Oslo, Norway (Holestöl and Palmström, 1996); and the Juktan hydropower station in Sweden (Rundgren and Martna, 1989).

in their formation and activity and may be studied to inform decision making. Albert and Barabási (2002) describe the statistical mechanics and dynamics of networks that at one time seemed random, and Watts (2004) summarizes findings about networks, network organizations, and the collective dynamics within networks than can, among other things, foster or inhibit information dissemination. There are extensive literature and computational analyses of analogous complex sociotechnical systems that can be applied to this discussion. For example Carley and others (2009) use quantitative analysis techniques to determine how learning occurs within networks that result in change, and Cataldo and others (2008) explore modeling how different types of software engineering decisions constrain other software engineering decisions and drive the need to coordinate activities.

For the sake of this discussion, networks are simplified into two categories: (1) technical networks involved in the design, construction, operation, and maintenance of underground space and (2) organizational networks of government agencies, private-sector entities, and community groups that pay for construction, endure disruptions, and benefit from completed underground facilities. Ideal decision making occurs with continuous interaction between these two networks during all phases of infrastructure life cycle. Identifying the right kind of information to share with the right agents within the right networks to facilitate change that promotes sustainability is difficult, and there is no singular methodology that will work in all urban systems, or possibly within a single urban system over time given the individual and dynamic nature of the networks.

Sustained support of infrastructure investment requires an understanding of how the press and public will perceive the project and associated activities, and how information can be transferred to them. The commitment of political leadership for the duration of project construction, operation, and maintenance is also needed. Public satisfaction with investment in infrastructure requires transparent communication including accurate representation of the value and risks of investment such as those associated with project cost and scheduling. Difficulties sustaining public support for investment decisions may lead to overpromising on design, analysis, and construction in order to get projects under way.[3] It may be possible to develop and use tools to raise the collective awareness in the community of the benefits and costs associated with underground infrastructure. For example, geotechnical databases have been developed for multiple communities around the world that can visually display the relationships between built infrastructure and the geologic environment (Reeves, 2010; Thompson, 2010). These may be applied for educational and planning purposes. As explored further in Chapter 5, the comparative assessment of sustainability for underground and surface space-use options requires that adequate data and case examples be

[3]For example, the multibillion-dollar "Stuttgart 21" project in Germany has generated much opposition by those who believe the project is overambitious and overpriced (Ward, 2010).

documented so that whole life cycle and triple bottom line impacts of competing options can be evaluated.

User Acceptance of Underground Institutional, Commercial, and Industrial Facilities

Institutional, commercial, and industrial underground facilities have come to be viewed differently by those who work or spend long periods in the facility than by those who choose to use the facility for shorter periods of time. Workers desire spaces that are as comfortable and safe as aboveground facilities. The lack of access to natural light, ventilation, and a spatial frame of reference (e.g., a view) is the most often cited detriment (Carmody and Sterling, 1993). Hence, worker acceptance may depend on the extent to which the facility is underground or windowless, and the type of environment expected for their work in a conventional facility. On the other hand, user acceptance revolves around convenience and safety perceptions, in addition to comfort. Both workers and users can be strongly influenced by quality of design, maintenance, operation, and security. Private and public stakeholder acceptance may be affected strongly by location and design—e.g., does placing all or part of a structure underground enhance its attributes in that location? If so, then costs and user concerns are weighed against the benefits of constructing the facility in that location. Hotels in the Washington, D.C., area, for example, often build several levels underground for parking, meeting, and ballroom spaces. Architectural height restrictions in the D.C. area[4] mean that the "windowed" space aboveground is at a premium. Underground space development is the result of codes in place to preserve the aboveground environment. The environment draws people to the area, and the well-designed and safe underground space draws usage (see Box 2.2).

Driving Forces

It is difficult to assess what driving forces are the most important in either advancing or hindering the development and use of underground facilities. Most large urban areas within the United States and around the world exhibit the growth of underground facilities as urban development intensifies. In this regard, one might conclude that no special policies or drivers need to be in place to cause development of the underground—it will happen as a natural result of land use, environmental pressures, and the need to upgrade transportation and utility services for a growing city. The downside of this laissez-faire approach is the chaotic development of the underground, project by project, even when it is well understood in principle that expanded underground uses will follow later. This

[4]DC ST § 6-601

BOX 2.2

The Arthur M. Sackler Gallery and the National Museum of African Art, Smithsonian Institution, Washington, D.C.

The Arthur M. Sackler Gallery (an Asian art museum) and National Museum of African Art at the Smithsonian Institution in Washington, D.C., are placed underground in the courtyard adjacent to the Smithsonian "Castle" on the National Mall. The function fits an underground structure well—the appearance of the iconic Smithsonian building is preserved, no open space on the National Mall is covered, and museum workers are already used to working in above-ground windowless buildings. High design quality and interesting gallery spaces provide an attractive environment for the public. The figure demonstrates use of space design, art, and natural lighting to create a dramatic, pleasant environment.

Figure A sculpture by Xu Bing occupies and can be viewed on each of the four levels of the Arthur M. Sackler Gallery in a beautifully designed space that makes clever use of skylights and artificial lighting. The sculpture as viewed looking down from upper level of four levels. Note the fountain at the bottom of the sculpture that reflects natural light from the skylight four stories above. Credit: Andrea S. Norris.

section examines some of the drivers that can either promote or inhibit development of an expanded and well-ordered underground environment.

Urban planners may plan the city in only two dimensions (with the use of height controls or floor-area ratios used to control building heights) and ignore the importance of the underground in major urban areas. Without federal, state, or municipal mission agencies with an overarching responsibility for the provision of urban infrastructure, separate agencies deal independently with issues related to transportation, housing and urban development, homeland security, and energy. Although the individual mandates of these agencies are important, a common approach to underground utility provision and urban planning of the underground is missing. Funding mechanisms for projects or research tend to focus on particular problems or solutions without much consideration of how the solution affects the system of systems in short and long terms. Local or national economic recessions that make investment in public facilities less urgent and less affordable exacerbate the problem, as do initially higher underground project development costs and the long timescales until project completion. Negative perceptions about the interior environments of underground facilities, confusing layouts and lack of reference to surface landmarks that inhibit easy wayfinding, and fears about personal safety in what may be perceived to be poorly designed and operated underground facilities may decrease public support of underground infrastructure.

However, there are many examples of successful underground infrastructure projects that lead to more sustainable societies. Development of some of these is facilitated by the governance structures and systems in place in these locations, some very different from those found in the United States. The strong policies for new infrastructure provision coupled with strong administrative controls for project implementation found in China, for example, would not necessarily be implementable in the United States. Policies that require and facilitate effective long-range planning of underground space use, as are found in locations such as Singapore or Helsinki, Finland, help those locations move closer to sustainability goals. Policies that enforce preservation of the surface environment while permitting facility expansion underground would provide a reason for moving more infrastructure underground where other incentives are not present. These could include building height restrictions coupled with the exclusion of underground space from floor area limitations, or prohibition of overhead utilities. Policies that increase the possibility to easily route infrastructure elements at depth beneath private land, such as Japan's Special Measures Act for Public Use of Deep Underground (Act no. 87 of 2000; see Konda, 2003) that gives public organizations prior rights to develop deep underground space, can help to avoid some of the legal barriers to broader, more versatile, and rapid development.

Underground engineering and construction is expensive, and construction costs are generally greater than for surface infrastructure. However, full assessment of lifecycle costs and benefits (see Chapter 5) may convince owners and planners that the initial greater investment is the better investment. Changes in policy as described above could lower some costs by, for example, streamlining some of the time-consuming processes related to permitting and rights of way. Other economic drivers are more practical in nature. Urban area and economic expansion may create a demand for new facilities and services, but surface land may not be available given increased density of urban development. Still other drivers may be human behavioral, for example, an insistence by the public for a better surface environment and improved public and private services, more awareness of quality-of-life approaches taken in different cities, or demand or response to better design quality in underground facilities, and better integration with the surface that removes negative perceptions about underground space.

Multiple circumstances and drivers accelerate or inhibit acceptance and development of the underground. Issues related to perceived negative perceptions and comfort of the underground are discussed further in Chapter 4, to better assessment of true economic, social, and environmental costs are discussed in Chapter 5, and to improved technologies that allow better understanding of construction and operational risks that increase costs are discussed in Chapter 6. A new approach to infrastructure planning and management that values the contributions of underground engineering to sustainable development is suggested with the committee conclusions in Chapter 7.

It is better not to consider the underground as a universal alternative to the surface—it isn't—but rather to give due consideration of the underground with respect to the long-term future and sustainability of an urban area. It is critical that future underground development options are not degraded by unplanned or unsuitable earlier uses, that policies and administrative structures provide the right guidance, that the public is fully engaged in developing a long-term vision for its community and community standards, and community expectations regarding how underground facilities will serve it are met.

CROSS-SYSTEMS INTERDEPENDENCIES

As underground use becomes more complex, it is evident that proper respect of the interplay between the surface and underground is necessary during all phases of infrastructure life cycle. Examples of the serious negative effects of poor management of surface or underground infrastructure are provided throughout this report but are not presented to indicate that such is the norm in engineering practice. Box 2.3, for example, demonstrates the effects of a load-bearing structural failure in the underground during construction that compromised surface facilities. During construction, infrastructure is often more susceptible to structural failure because soils may not be fully stabilized until construction is complete. The stability of surface infrastructure is dependent on the stability of the subsurface. Numerous other interdependencies are less obvious. Many of these interdependencies may be critically important to national security.

The Presidential Commission on Critical Infrastructure Protection defined infrastructure systems vital to our country (PCCIP, 1997) and prospectively looked at critical infrastructure as the subject of planned measures to protect assets from damage or destruction. Sustaining our nation and way of life were considered dependent on the continued, uninterrupted services of these infra-

BOX 2.3

Failure of a Metro Line in Cologne, Germany, 2009

Construction of Cologne, Germany's 3.8 km North-South Metro Line (with tunnels for seven underground stations) began in 2004 and was completed in summer 2008. Construction of the stations, two emergency shafts, and an underground turn-off (28 m deep) included the use of cut-and-cover and mining methods with ground freezing. On March 3, 2009, Cologne's seven-story (five above ground and two below) Historical Archive building, located by the open pit of the underground turnoff, collapsed along with buildings, including homes, located on either side (see Figure 1). Caused by the inrush of ground water and ground material, the collapse resulted in loss of life and extensive damage to the voluminous and valuable historical records of the City of Cologne, the surrounding region, and Germany (Haack, 2009; see Figure 2). Debris and soil were deposited in the building, the temporary steel ceiling and the dewatering system were damaged, concrete was cracked, and surrounding soil was loosened and displaced. There is speculation that the collapse was due to local separation caused by the removal of soil by the dewatering system or a failure of the diaphragm wall structure (Manderfeld, 2010).

FIGURE 1 Collapse of multiple buildings resulting from excavation collapse. SOURCE: AP Images.

structure systems. More than a decade later and after the attacks of September 11, 2001 (9/11), the list of infrastructure defined as critical by the PCCIP still applies (NRC, 2002), but with greater urgency and entailing more issues, hazards, and levels of protection. Because of the events of 9/11 and significant regional days-long power outages (for example, see Minkel, 2008), the interdependency of infrastructure systems has been elevated to a matter of national concern.

Perhaps in part because of attention provoked by the 9/11 attacks, critical infrastructure networks are now recognized as interdependent systems (NRC, 2002). They include systems that provide potable water, wastewater and stormwater collection and disposal, electric power, fuel distribution, telecommunications, and digital television and Internet connectivity and communications

Two people in collapsed houses died. Forty-five others (30 in the Archive and 15 residents) were rescued (Haack, 2009). Public anger and a threat that an angry mob would demand abandonment of the entire project, intense speculation about what had occurred, and a "vacuum of official details" were reported after the accident (Wallis, 2009). Initial political responses were that the construction plan in Cologne (or in any densely built-up town) should not have been approved (Haack, 2009). According to a September 2010 engineering report, the cause was still under investigation by the public attorney's office. Independent experts nominated by the court are reviewing the incident (Manderfeld, 2010). The reverberations of the Cologne accident extended to Amsterdam where a metro line was being planned for an area with similar geologic characteristics. Twice in 2008, there was damage from leaking in the concrete wall of a construction pit for a future metro station on the Vijzelgracht, Amsterdam, and, as a result, neighboring 17th-century weavers' houses became flooded and unfit for habitation (van Outeren, 2009).

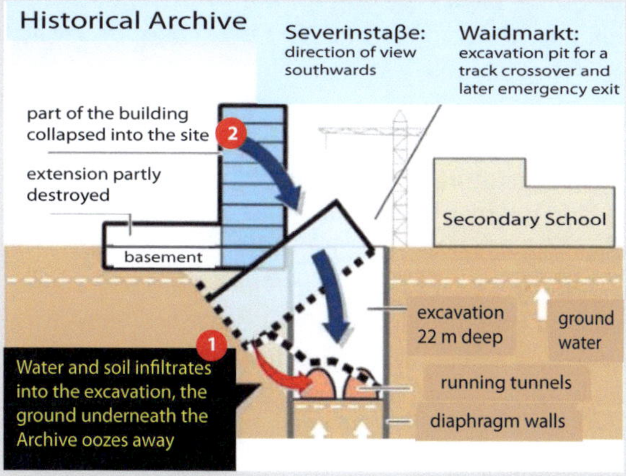

FIGURE 2 Diagram illustrating the assumed cause of the accident. SOURCE: Haack, 2009. Reprinted with permission of Alfred Haack.

(including orbiting satellite assets), as well as transportation systems such as roads, highways, bridges, tunnels, transit and railroad facilities, and airports and harbors (Peerenboom, 2001). Other infrastructure systems also provide emergency services, living and working spaces, churches and places of assembly, hospitals and schools, parks and recreation areas, open spaces, and other facilities (NRC, 2002).

As systems, they are characterized in part by a complexity related to the fact that they are owned and controlled by numerous individuals, partnerships and corporations, and local, state, and national governments. This complex ownership model leads to confusion regarding, for example, responsibility for funding and performing essential periodic inspections, maintenance, and repair of individual infrastructure elements or systems. There is no clear or consistent understanding of who does what, under what guidelines and budgets, and under what terms.

Cascading Failures of Systems

Interdependencies among infrastructure systems are often not fully understood (Little, 2005). Failure in one element of a system can cause disruptions in one or many other systems, and failure of underground systems can occur as a result of failure of systems on the surface. Disruptions can spread to systems in other cities, states, and countries. For example, cascading failure of interdependent underground infrastructure occurred as a result of the 9/11 attacks on surface infrastructure in New York City. Water main breaks flooded rail tunnels, a commuter station, and a facility that housed all cables for what has been described as the world's largest telecommunication node. Trading on the New York Stock Exchange ceased for six days as a result of failure of communication infrastructure. International financial stability therefore was linked to a water main rupture in one location (O'Rourke, 2007).

Communication systems failure can result in the cascading failures of electric-powered plants, systems, and equipment. Electric power systems are more often remotely controlled from a central operations station, by wireless or leased telephone lines, the Internet, or by supervisory control and data acquisition (SCADA) systems. SCADA systems typically use open architecture software without security protection, making them vulnerable to hackers. Access to a SCADA system could provide opportunity to cause problems with system functionality including overloading a transmission grid (NRC, 2002). SCADA systems can also malfunction when electric power fluctuates or becomes unstable, as was demonstrated by the 2010 natural gas pipeline explosion in San Bruno, California (NTSB, 2011a).

Another example of cascading failure of interdependent infrastructure occurred in August 2003 when an overloaded Ohio utility electric transmission line faulted, shutting down a portion of the transmission grid, leading to failure of the electrical transmission network and the blackout of eight states in the Northeastern United States and southeastern Canada. Fifty million people lost power for up to two days (Minkel, 2008). Cleveland, Ohio, did not have power to pump public water to 1.5 million citizens (Little, 2005), and similar situations were reported elsewhere. Loss of power shut down traffic controls and street lighting, making road and highway travel hazardous, particularly at night. The effects of the failure were far reaching. Refrigeration for food was impossible, and emergency measures were needed, for example, to protect children's milk supplies (PSEPC, 2006). Service stations could not pump fuel, and people abandoned vehicles wherever they ran out of gas. The rapid cascading effects on other critical services were also observed (Minkel, 2008).

Cascading failure of interdependent infrastructure may result when existing underground infrastructure is disturbed by the installation or repair of other services. "Call before you dig" (CBYD) laws[5] have been put in place in many areas to reduce this likelihood. More attention needs to be directed to the buried resources that, in many cases, are just below the surface. It is in the interest of property owners and managers to know the location, condition, and state of repair of infrastructure elements that service their properties. Local governments, under public safety and health mandates, have a role in assuring that inspections and maintenance of lifeline infrastructure occurs and is documented and available. State governments have similar responsibilities for electrical power grids, transmission pipelines, and potable water supply systems. Less labor-intensive means of mapping underground utilities, performing and reporting essential lifeline service inspections, and understanding the implications of their interconnections on local users could lead to a better systems approach to planning, construction, operation, and maintenance over the long term to increase the life of critical infrastructure and avoid cascading systems failures. The committee suggests potential research in these areas in Chapter 7.

CONSEQUENCES OF INCOMPLETE PLANNING

The study committee began with an assumption that sustainable development is dependent on the ability of planners to consider future needs. The useful life of critical infrastructure is dependent on the service being installed and determined during design and materials specification processes. Buried utility services are expected to operate for 50 years; transit and sewer tunnels and structures for 100 years. It is often difficult to predict how best to accommodate long-term operation and maintenance of the infrastructure while simultaneously accommodating growing or changing populations, changing infrastructure needs, and new technologies. It is especially difficult to predict what may be the societal needs of infrastructure in 50 to 100 years. Practical methods for determining remaining useful life of utilities and services are needed.

The next sections highlight issues that result from poor or incomplete planning and how these issues relate to those associated with aging infrastructure and the choice of building foundations.

Aging Not-So-Gracefully while Keeping up with Demand

There were approximately 76 million people in the United States at the beginning of the 20th century, and there are approximately 310 million people today (U.S. Census Bureau, 2012). The U.S. Census Bureau estimates the U.S.

[5]For example, see Oregon Law OAR 952-001-0010 through OAR 952-001-0090, available at arcweb.sos.state.or.us/rules/OARS_900/OAR_952/952_001.html. See also www.callbeforeyoudig.org/law.htm (accessed November 11, 2010).

BOX 2.4

Failure of a Water Main

The 2008 failure of a 66-inch pre-stressed concrete water main resulted in the need for rescue of nine people stranded dangerously in their cars while approximately 150,000 gallons of water per minute rushed down a major street near Potomac, Maryland (Morse and Shaver, 2008). The pipe, 15 feet below the surface, was put into service in 1964. A forensic investigation indicated that pipe corrosion and weakening was caused by the installation of the pipe directly on rock (WSSC, 2009). The pipe was last internally inspected in 1998, but internal inspections do not normally expose the external chemical-based corrosion that occurred.

A 50-by-30–foot hole was created by the force of the rupture, several large trees and a utility pole were downed, and a portion of the road was destroyed. Area schools and roads were temporarily closed. As a result of this event, the responsible agency immediately implemented a real-time, active monitoring program for a majority of its large diameter water main system.

population will be approximately 439 million by 2050 (U.S. Census Bureau, 2008), representing a 42 percent growth in the next 40 years. The public expects delivery of a certain quality of life through physical infrastructure, and such growth will create financial and physical pressures to enlarge all infrastructure systems while concurrently identifying ways to extend the useful life and reliability of existing systems. Different and even greater demands on infrastructure will be likely as technologies evolve and new technologies are developed and their delivery becomes expected.[6] Infrastructure interdependencies will likely become even more complex, and, as infrastructure systems age, the system of systems is likely to become less reliable. This is not a good scenario for sustainability.

It is reported that a significant portion of the underground infrastructure in the United States is at or has exceeded its projected useful life (USNCTT, 1989; ASCE, 2009). Responsible agencies seek effective ways to stretch dwindling budgets and capital expenditures to address issues associated with aging infrastructure, but a gap exists between appropriated funds and expenditures necessary for infrastructure renewal. In 2002, the U.S Environmental Protection Agency (EPA) forecasted an $8 billion annual gap over a 20-year period (2000-2019) for the nation's aging water infrastructure alone (EPA, 2002). The American Society of Civil Engineers (ASCE) estimated that leaking water pipes result in the loss of 7 billion gallons per day, nationwide, of clean drinking water (ASCE, 2009). Further, the ASCE described that deteriorated wastewater pipelines leak billions of gallons of sewage into the nation's waterways each year. Personal and economic safety and health may be put at greater risk by an inability to mitigate projected

[6] For example, Internet access, unheard of just a few decades ago, is considered by many to be a "fundamental right." See BBC, 2010.

BOX 2.5
Pipeline Failures in San Bruno, California, and Carmichael, Mississippi

Recent pipeline failures in San Bruno, California, and Carmichael, Mississippi, demonstrate the uncommon but significant risk to surface infrastructure, especially in highly populated areas. The San Bruno, California, natural gas pipeline explosion and fire in 2010 illustrates the potential risks associated with a buried gas pipeline. The line was installed in 1956 beneath land that was subsequently developed into a thriving residential neighborhood. This pipeline relied on a dedicated SCADA system for control of gas flow and pressure. Eight people were fatally injured, and more than 50 residences were damaged or destroyed as a result of the explosion of the 30-inch-diameter steel gas pipeline. A preliminary report from the National Transportation Safety Board (NTSB) indicated the rupture occurred following a power malfunction of the electrical line feeding the SCADA system and resulted in an increase in pressure on the line (NTSB, 2011a). All of these factors point toward finding better means for developing, operating, and maintaining our infrastructure systems.

In 2007, the rupture of a pipeline transporting liquid propane in rural Mississippi released more than 10,000 barrels (approximately 430,000 gallons) of propane. The propane formed a gas cloud and ignited, creating a large fireball that resulted in two fatalities, seven injuries, and four destroyed houses (NTSB, 2009). About 70 acres of grassland and woodland were burned, and more than $3 million of property damages were claimed lost by the pipeline company. The NTSB determined that among the several safety issues contributing to the incident was the inadequacy of regulation and oversight exercised by the Pipeline and Hazardous Materials Safety Administration of pipeline operators' public education and emergency responder outreach programs (NTSB, 2009)

Such events are relatively uncommon. The NTSB lists 17 significant pipeline incidents investigated in the United States in the past 10 years (NTSB, 2011b).

infrastructure system failures that directly and physically endanger citizens (see Boxes 2.4 and 2.5 for examples).

Building Foundations and Future Underground Use

Building foundations constitute a major use of urban underground space, provide necessary building support, and can add value and space to properties. However, building foundations are rarely designed with thought to how the space under or surrounding the foundation may be used in the future. Deep pile foundations of some structures, for example, may make it more difficult to accommodate infrastructure such as transit and road tunnels that have significant horizontal and vertical alignment restrictions. Some foundation designs may restrict public rights-of-way. The installation of horizontal soil anchors or "nails" provides lateral support for deep foundations and walls against the pressure of surrounding soil and groundwater, but requires the placement of tiebacks in holes on the sides of an excavation that can extend 30 to 35 feet into adjacent soil or rock. Although tiebacks generally serve no structural post-construction function, they often are left in place and may compromise other uses of the underground in their locations.

Foundation design and construction could include sustainable practices such as the use of removable anchors, if feasible. A longer term approach to foundation design might include designing foundations so that they are more readily reusable or repurposed once the surface infrastructure outlives its useful life. Current practices for reconstruction often include demolition of surface and foundation structures when new construction occurs. A visionary approach to foundation design requires consideration of urban sustainability holistically. It accounts for the collective impact of individual design and construction decisions on future use of the urban underground. Designs may take into account long-term planning for the urban area as a whole, for example, avoiding specific designs in an area zoned for future underground transportation. This approach will be more successful when the urban underground is incorporated into urban growth plans as part of a functioning and evolving system of systems. Optimal planning may sometimes call for preserving the underground for future use.

Institutional Management of Underground Space

As has been described, decisions related to individual underground infrastructural elements are seldom made using a systems management approach in which above- and belowground infrastructure, combined, comprise an integrated system. Governance and institutional management of urban underground space that guides decision making in the United States is fragmented at best, and nonexistent at worst. Public policies that govern urban underground use, with few exceptions, are not well formed. The primary focus of urban planning is the provision of services under the constraints of available surface and air rights and resource development. A great challenge to governance is that ownership of underground utilities, services, and structures is vested in a variety of public and private parties. This and the lack of frameworks for valuation of underground space by municipalities are among issues that frustrate better urban underground planning and management. Municipalities typically allow subsurface operations in their jurisdiction through permitting processes, but lack authority to regulate. Permission to cut into an existing street for any purpose, for example, may require a permit, but the permit typically does not include conditions specific to the utility or service being installed. Further, submission of as-built records of installations may not be required.

Comprehensive mapping of the locations of buried utilities and services are rarely available, and as-built records of existing underground facilities may be either publicly unavailable or inaccurate. It is therefore difficult to plan new installations without disturbing existing buried services. According to the Common Ground Alliance,[7] a utility is hit or damaged in the United States every 60 seconds (Landes, 2008). Available technologies to detect and map underground infrastructure to minimize striking and damaging their systems during construction activities are not employed often enough.

A national 811 number—the "call before you dig" line[8]—was launched in 2007 to reach 62 call centers connected with parties that have buried services in their coverage areas. This is a first step in developing institutional management of urban underground space. A positive outcome of CBYD is the sharing of utility and services data by interested parties. The governance gap can begin to close when public policy requires accounting for the use and optimization of underground urban space for the benefit of the people, the economy, and sustainable development.

PLANNING AND GOVERNANCE FOR SUSTAINABILITY

Some cities around the world have made greater progress planning and governing underground space. Helsinki, Finland, a notable example, has identified and protected its prime near-surface rock resources and has developed deep common utility tunnels that limit interference with shallower, people-oriented underground infrastructure, such as that used for transit, pedestrian connections, and parking. This strategy moves away from the more common practice of placing utilities directly beneath the surface. Montreal, Canada, has established the framework by which a largely private network of underground pedestrian connections in the downtown area has turned a northern-climate city into an extensive indoor city that is comfortable and accessible in the harshest winter weather. Perhaps the most ambitious underground planning at the time of writing is being undertaken by the City State of Singapore. The extreme shortage of land and natural resources of this island nation makes use of underground space an important component of overall planning (Hulme and Zhao, 1999). Effective underground space use in Singapore preserves surface space for other uses, including recreation.

The European Construction Technology Platform promotes the concept of a multidimensional city in which people move vertically above and below ground as well as horizontally (ECTP, 2005). Box 2.6 demonstrates European Union

---

[7]The Common Ground Alliance is an association dedicated to public safety through damage prevention practices. For more information see http://www.commongroundalliance.com/Template.cfm?Section=About_CGA.

[8]For more information see http://www.call811.com.

BOX 2.6

A Strategic Research Agenda for the European Construction Sector

The European Construction Technology Platform (ECTP)[a] developed a strategic research agenda for the underground construction sector for the next 25 years that takes into account innovations driven by the market and long-term societal visions (ECTP, 2005). According to the ECTP, future models for urban planning must incorporate new ways to think about underground space and construction concepts so that underground space use can be expanded downward as far as imagination and technologies will allow. Underground infrastructure will be more appealing when directly and conveniently linked to improved surface space and to high-capacity transportation systems that are efficient alternatives to surface transport. The figure is a schematic of what the platform termed a multidimensional city.

The ECTP suggests that all aspects of construction (e.g., organization of supply chains, contractual arrangements, service industries, underground architecture, specialized vehicles, technologies for excavation, social business, and the safety and security industry) must be reviewed and revamped to improve work within an underground environment and to provide supervision and protect against hazards. The ECTP's research agenda includes a vision and short- and long-term research priorities intended to meet the needs of clients (e.g., through efficient use of the underground and improving our understanding and ability to control the ground itself), allow cities to become sustainable (reducing resource consumption, environmental and anthropogenic impacts, improving safety and security, and enhancing the quality of life), and cause a transformation in the construction sector itself (through increased competiveness, a new knowledge-based construction process driven by clients, information and communication technologies and automation, state-of-the-art construction materials, and attractive work environments) (ECTP, 2005).

[a]See http://www.ectp.org/ (accessed October 6, 2011).

recognition of the importance of urban underground space and its vision for the impact on city livability possible with integrated space resource planning.

Growing urban populations have resulted in development of marginal lands (i.e., weak and soft soils) and underused industrial and commercial facilities and associated poor environmental conditions (e.g., pollution, hazardous waste, and contaminated ground). Underground development may also encroach on marginal lands, and developers and contractors must deal with issues such as hazardous waste removal or remediation. This warrants thoughtful and extended consideration by owners, urban planners, developers, and the public about the geotechnical and geo-environmental issues related to all urban construction, about underground space development specifically, and about the explicit valuation of underground space as a resource.

Effective planning and infrastructure investment decisions require that relevant administrators and planners accept the need and responsibility for integrated

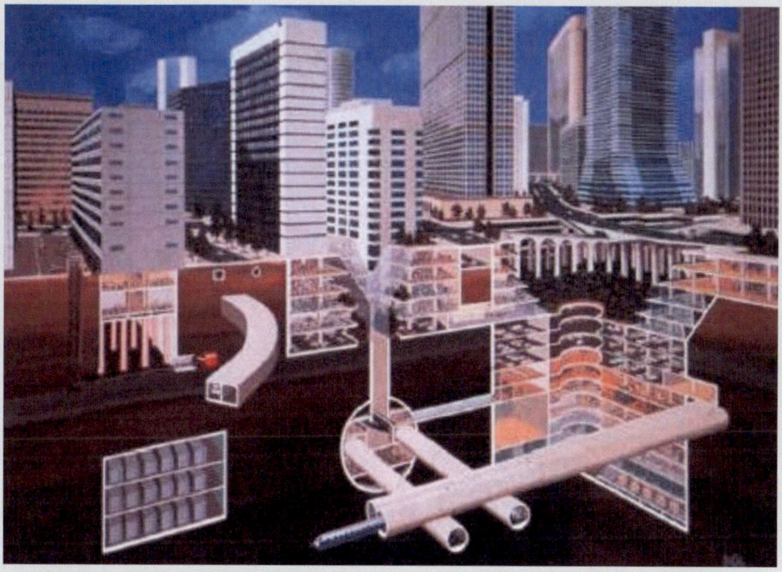

FIGURE A new multidimensional city as envisioned by the Focus Area of Underground Construction of the European Construction Technology Platform. SOURCE: ECTP, 2005, p. 9.

long-term planning, and information archiving is needed that ensures information resources are available in a useful form. This implies that both reliable geological and three-dimensional records from multiple sources of existing structures in the underground need to be developed, registered to a common spatial reference, and maintained. Visualization for underground planning is needed particularly in complicated geologies, with significant topographic variations, and when multiple levels of underground facilities are considered (Reeves, 2010). The ability to archive, search, manage, and display complex three-dimensional databases at appropriate degrees of complexity for planning and detailed design tasks would greatly aid the ability to effectively plan urban underground space use. Some aspects of the databases and software needed to undertake this task exist, but many complications remain in terms of permission to access detailed private utility data and to manage the uncertainty and varying quality of available data (Reeves, 2010). This discussion is continued in more detail in Chapter 6.

To consider explicit cost-benefits in infrastructure decisions, it is important to establish a methodology to quantify the value of subsurface space opportunities as a resource in urban environments. This would allow comparison of the value of underground space on a par with other urban resources, for example linked to an increasing market value of surface land property. Value for future uses would also encompass the fact that the nature of previous use (e.g., existing infrastructure) can force new infrastructure systems to be placed in increasingly difficult ground conditions, presenting problems for engineers and constructors, and creating additional difficulties related to scheduling and cost control. Effective planning and governance can help efficiently optimize use of underground resources and obtain the most value from the underground resource for the long term. Governance approaches include zoning subsurface vertical and horizontal space, reserving corridors for major transportation systems, and coordinating utility space use requirements in the public rights-of-way.

LONG-TERM MANAGEMENT OF THE UNDERGROUND

There are many fabled successes in underground infrastructure (e.g., the New York City and Boston subway systems) and more recent successes in underground infrastructure development—the Washington, D.C., Metro, the Metropolitan Atlanta Rapid Transit Authority, and the Chicago Transit Authority. The record of accomplishments extends to the creation of underground utility systems. However, the legacy of more than a century of abandoned or unmapped subsurface infrastructure also presents great problems (Sterling et al., 2009). Positions of abandoned utilities, foundations, tanks, and construction or demolition debris are not recorded or their records discarded. Positions of active utilities can be uncertain or improperly recorded. This situation is not unique to the United States, and some parts of the world have an even longer legacy of abandoned buried infrastructure. A reasonable step toward sustainable planning practices would be the development of a geographic information system database with information about locations of underground infrastructure and artifacts. A 10-year research program is under way in the United Kingdom to develop a prototype multi-sensor ground penetration radar tool that would locate and map buried utilities and services. Three-dimensional maps would then be made in conjunction with the British Geological Survey.[9] More reliably documenting all things underground in a searchable database system that includes tools for visualization—and documenting other unrecorded services encountered during underground construction—would vastly improve the ability of planners to maximize

[9]See http://www.mappingtheunderworld.ac.uk/ (accessed September 15, 2011).

the use of the underground while minimizing the cost of building and maintaining underground infrastructure.

From 1972 to 1994, the U.S. National Committee on Tunneling Technology within the National Research Council served as the national organization to stimulate advances in tunneling technology and subsurface use (see Appendix C). The committee did not have oversight responsibilities, but it did serve to shape technology, practice, and education and training. Membership included representatives from government, industry, and academe. Its purpose was to promote coordination of activities, including assessment, research, development, education, training, and dissemination of information. It also served as the U.S. adherent to the International Tunneling Association. There has been no similar body since, and in 2012 no official bodies in the United States carried the responsibility of overseeing and approving use of the underground to manage it in the most sustainable manner. The lack of planning for systematic and sustainable use of the underground results in significant added costs and schedule difficulties as new services are installed in very congested urban underground space. The United States envisions installation of High Speed Railroad (HSR) systems, some of them underground, as built in other parts of the world. Grade-separated freight movement systems (for example, railroad tracks and truck roadways) could also be placed underground as part of sustainable urban development. The cost of these or any future underground infrastructure in urban settings will increase because of the inability to plan effectively around existing infrastructure. Research opportunities to develop a framework and management approach to planning, documenting existing conditions, setting land use requirements, and issuing permits for approved uses of the urban underground can be found in Chapter 7.

REFERENCES

Albert, R., and A.-L. Barabási. 2002. Statistical mechanics of complex networks. Review of Modern Physics. 74(1):47-97 [online]. Available: http://www.barabasilab.com/pubs/CCNR-ALB_Publications/200201-30_RevModernPhys-StatisticalMech/200201-30_RevModernPhys-StatisticalMech.pdf (accessed November 8, 2012).

ASCE (American Society of Civil Engineers). 2009. 2009 Report Card for America's Infrastructure. American Society of Civil Engineers [online]. Available: http://www.infrastructurereportcard.org/sites/default/files/RC2009_full_report.pdf (accessed May 1, 2012).

BBC. 2010. Internet access is 'a fundamental right'. BBC News: March 8, 2010 [online]. Available: http://news.bbc.co.uk/2/hi/8548190.stm (accessed March 1, 2011).

Broch, E. 2006. Use of Rock Cavern in Urban Areas in Norway [online]. Available: http://www.ita-aites.org/fileadmin/filemounts/general/pdf/ItaAssociation/Organisation/Members/MemberNations/Egypt/PapersSharm2006/kn05.pdf (accessed May 2, 2012).

Brongers, M.P.H. 2002. Appendix K: Drinking Water and Sewer Systems. Cost of Corrosion, March 2002 [online]. Available: http://www.corrosioncost.com/pdf/water.pdf (as cited in Sterling et al., 2009).

BTS (Bureau of Transportation Statistics). 2004. The Transportation Network: 2004 [online]. Available: http://www.bts.gov/publications/pocket_guide_to_transportation/2006/html/table_01.html (as cited in Sterling et al., 2009).

Butts, C.T. 2009. Revisiting the foundations of network analysis. Science 325:414-416.

Carley, K.M., M.K. Martin, B.R. Hirshman. 2009. The etiology of social change. Topics in Cognitive Science 1(4):621-650. Available: http://onlinelibrary.wiley.com/doi/10.1111/j.1756-8765.2009.01037.x/pdf (accessed November 6, 2012).

Carmody, J., and R.L. Sterling. 1993. Underground Space Design: A Guide to Subsurface Utilization and Design for People in Underground Spaces. New York: Van Nostrand Reinhold.

Cataldo, M., J. D. Herbsleb, and K.M. Carley. 2008. Socio-Technical Congruence: A Framework for Assessing the Impact of Technical and Work Dependencies on Software Development Productivity. International Symposium on Empirical Software Engineering and Measurement Proceedings of the Second ACM-IEEE [online]. Available: http://herbsleb.org/web-pubs/pdfs/cataldo-socio-2008.pdf (accessed November 6, 2012).

DOE (U.S. Department of Energy). 2006. Transmission, Distribution & Storage. Oct. 2006 [online]. Available: http://www.fossil.energy.gov/programs/oilgas/delivery/index.html (as cited in Sterling et al., 2009).

ECTP (European Construction Technology Platform). 2005. Strategic Research Agenda for the European Underground Construction Sector, Draft, October 2005. European Construction Technology Platform [online]. Available: http://www.ectp.org/documentation/FA-Underground-Constructions_SRA-VISION2030-23Nov05.pdf (accessed May 2, 2012).

EPA (U.S. Environmental Protection Agency). 2002. The Clean Water and Drinking Water Infrastructure Gap Analysis. EPA-816-R-02-020. Office of Water, U.S. Environmental Protection Agency [online]. Available: http://www.epa.gov/ogwdw/gapreport.pdf (accessed November 11, 2010).

EPA . 2006. Emerging Technologies for Conveyance: New Installations and Rehabilitation Methods. EPA 832-R-06-004. July 2006 (as cited in Sterling et al., 2009).

EPA . 2007. P. 13 in Distribution System Inventory, Integrity and Water Quality, Jan. 2007 (as cited in Sterling et al., 2009).

Farrimond, M. 2004. The ESWRAC Initiative. Proceedings, ISTT No-Dig 2004 Conference. November 15-17, 2004, Hamburg, Germany [online]. Available: http://www.eswrac.org/download/1/Farrimond.pdf (accessed May 1, 2012).

FCC (U.S. Federal Communications Commission). 2006. Table I.A: Outside Plant Statistics—Cable and Wire Facilities. FCC Report 43-08: ARMIS Operating Data Report (as cited in Sterling et al., 2009).

FHWA (Federal Highway Administration). 2008. To Move Forward with FHWA Initiatives. The Environmental Quarterly 4(3) [online]. Available: http://www.fhwa.dot.gov/resourcecenter/teams/environment/vol4iss3.cfm (accessed May 1, 2012).

Glass, R.J., T.J. Brown, A.L. Ames, J.M. Linebarger, W.E. Beyeler, S.L. Maffitt, N.S. Brodsky, P.D. Finley, S.H. Conrad, T.F. Corbet, Jr., R.J. Detry, P.S. Downes, M.A. Ehlen, P.G. Kaplan, A. Kelic, M. Mitchell, R.A. Mitchell, T.W. Moore, A.V. Outkin, T.-T. Quach, G.E. Reedy, E.D. Russell, Sh. J. Starks, D.J. Sunderland, V.N. Vargas, S. J. Verzi, A.A. Zagonel, E.D. Vugrin. 2011. Pheonix: Complex Adaptive System of Systems (CASoS) Engineering Version 1.0. Sandia Report SAND 2011-3446 [online]. Albuquerque, NM: Sandia National Laboratories. Available: http://www.sandia.gov/CasosEngineering/docs/Phoenix_1.0_12_1_2011_%203446.pdf (accessed November 7, 2012).

Glass, R.J., A.L. Ames, W.A. Stubblefield, S.H. Conrad, S.L. Maffitt, L.A. Malczynski, D.G. Wilson, J.J. Carlson, G.A. Backus, M.A. Ehlen, K.B., Vanderveen, D. Engi. 2008. A Roadmap for the Complex Adaptive Systems of Systems (CASoS) Engineering Initiative. SAND Report 2008-4651 [online]. Albuquerque, NM: Sandia National Laboratories. Available: http://www.sandia.gov/CasosEngineering/docs/CASoSEngineeringRoadmap_09.22.08.pdf (accessed October 26, 2012).

Guo, D., P.P. Nelson, J. Xie, Z. Chen. In press. Vertical Planning of Urban Underground Space Use in China. 13th World Conference of the Associated Research Centers for Urban Underground Space, Singapore, November 7-9, 2012.

Haack, A. 2009. Construction of the North-South-Metro Line in Cologne and the Accident on March 3rd, 2009. STUVA (Research Association for Underground Transportation Facilities), Cologne, Germany [online]. Available: http://management.kochi-tech.ac.jp/ssms_papers/sms10_194_Alfred%20Haak100126(re-submit).pdf (accessed April 29, 2011).

Holestöl, K., and A. Palmström. 1996. Extensive use of the underground in Oslo is giving a better city. Pp. 103-114 in Proceedings of International Conference on Tunnels for the Third Millennium, Prievidza, Slovakia [online]. Available: http://www.rockmass.net/ap/56_Holestol&Palmstrom_on_Use_of_Oslo_underground.pdf(accessed May 2, 2012).

Hulme, T.W., and J. Zhao. 1999. Underground space development in Singapore: The past, present and future. Tunnelling and Underground Space Technology. 14(4):407.

Konda, T. 2003. Reclaiming the underground space of large cities in Japan. In (Re)Claiming the Underground Space. J. Saveaur. Swets & Seitlinger B. V.: Lisse, The Netherlands.

Landes, S. 2008. Underground damage occurs once every 60 seconds: Statistics you can't afford to ignore. Underground Construction (Jan):34-36 [online]. Available: http://www.rhinomarkers.com/DP_Info/Article_PDFs/DIRT_Underground_Construction_01_2008.pdf (accessed May 1, 2012).

Little, R.G. 2005. Tending the infrastructure commons: Ensuring the sustainability of our vital systems. Structure and Infrastructure Engineering. 1(4):263-270.

Manderfeld, J. 2010. Cologne's Tunnel Tube Disaster, 03.03.2009. Presentation at Annual IMIA Conference, September 14, 2010, Berlin [online]. Available: http://www.imia.com/downloads/guest_presentations/GP19_2010.pdf (accessed April 29, 2011).

Minkel, J.R. 2008. The 2003 Northeast Blackout - Five years later. Scientific American, August 13, 2008 [online]. Available: http://www.scientificamerican.com/article.cfm?id=2003-blackout-five-years-later (accessed April 8, 2011).

Morse, D., and K. Shaver. 2008. Water main break forces dramatic rescue of nine. The Washington Post, December 24, 2008 [online]. Available: http://www.washingtonpost.com/wp-dyn/content/article/2008/12/23/AR2008122302853.html (accessed April 7, 2011).

NERC (North American Electric Reliability Council). 2006. Long-Term Reliability Assessment: The Reliability of the Bulk of Power Systems in North America. Oct. 2006 [online]. Available: http://www.nerc.com/files/LTRA2006.pdf (as cited in Sterling et al., 2009).

Nordmark, A. 2002. Overview on survey of water installations underground: Underground water-conveyance and storage facilities. Tunnelling and Underground Space Technology. 17:163-178.

NRC (National Research Council). 2002. Making the Nation Safer: The Role of Science and Technology in Countering Terrorism. Washington, DC: The National Academies Press.

NRC. 2011. An Assessment of the Science Proposed for the Deep Underground Science and Engineering Laboratory (DUSEL). Washington, DC: National Academies Press.

NTSB (National Transportation Safety Board). 2009. Rupture of Hazardous Liquid Pipeline with Release and Ignition of Propane, Carmichael, Mississippi, November 1, 2007 [online]. Available: http://www.ntsb.gov/news/events/2009/Carmichael_ms/index.html (accessed May 2, 2012).

NTSB. 2011a. Pacific Gas and Electric Company: Natural Gas Transmission Pipeline Rupture and Fire, San Bruno, California, September 9, 2010. Pipeline Accident Report No. NTSB/PAR-11/01. National Transportation Safety Board [online]. Available: http://www.ntsb.gov/doclib/reports/2011/PAR1101.pdf (accessed May 1, 2012).

NTSB. 2011b. Pipeline Accident Reports [online]. Available: http://www.ntsb.gov/investigations/reports_pipeline.html (accessed May 8, 2012).

O'Rourke, T.D. 2007. Critical infrastructure, interdependencies, and resilience. The Bridge 37(1):22-29 [online]. Available: http://www.nae.edu/File.aspx?id=7405 (accessed February 15, 2011).

Parker, H.W. 2004. Underground Space: Good for Sustainable Development, and Vice Versa. International Tunnelling Association (ITA) Open Session World Tunnel Congress, May 2004, Singapore [online]. Available: http://www.ita-aites.org/fileadmin/filemounts/general/pdf/ItaAssociation/ITAEvents/OpenSessions/HParker.pdf (accessed April 30, 2012).

PCCIP (President's Commission on Critical Infrastructure Protection). 1997. Critical Foundations: Protecting America's Infrastructures [online]. Available: http://www.fas.org/sgp/library/pccip.pdf (accessed November 23, 2010).

Peerenboom, J. 2001. Infrastructure Interdependencies: Overview of Concepts and Terminology. National Science Foundation Workshop, June 2001 [online]. Available: http://we-partner.org/onionbelt/wp-content/uploads/2011/06/peerenboom_pdf.pdf [accessed May 7, 2012].

PHMSA (Pipeline and Hazardous Materials Safety Administration). 2003. Frequently Asked Questions: General Pipeline FAQs [online]. Available: http://www.phmsa.dot.gov/media/pipeline_qa.html (as cited in Sterling et al., 2009).

PHMSA. 2005. Gas Distribution Systems Annual Report Data File Fields, Distribution Annual Report Form RSPA F 7100.1-1, Dec. 2005 (as cited in Sterling et al., 2009).

Pipeline 101. 2001. Overview: How Many Pipelines Are There? [online]. Available: http://www.pipeline101.com/Overview/energy-pl.html (as cited in Sterling et al., 2009).

PSEPC (Public Safety and Emergency Preparedness Canada). 2006. Incident Analysis: Ontario–U.S. Power Outage—Impacts on Critical Infrastructure. IA06-002. August 2006 [online]. Available: http://www.publicsafety.gc.ca/prg/em/_fl/ont-us-power-e.pdf [accessed May 3, 2012].

Reeves, H. 2010. Attributed 3D Geologic Models in the UK and Their Application to Planning and the Sustainable Development of Underground Space. Presentation at the First Meeting on Underground Engineering for Sustainable Underground Development, June 1, 2010, Washington, DC.

Rundgren, L., and J. Martna. 1989. Underground hydropower projects in Sweden. Tunnelling and Underground Space Technology. 4(2):131-137.

Schewe, P.F. 2007. The Grid: A Journey through the Heart of Our Electrified World. Washington, DC: Joseph Henry Press.

Sterling, R.L., J. Anspach, E. Allouche, and J. Simicevic. 2009. Encouraging Innovation in Locating and Characterizing Underground Utilities. Report S2-R01-RW. Transportation Research Board [online]. Available: http://onlinepubs.trb.org/onlinepubs/shrp2/shrp2_S2-R01-RW.pdf (accessed July 10, 2012).

Sterling, R., H. Admiraal, N. Bobylev, H. Parker, J.P. Godard, I. Vähäaho, C.D.F. Rogers, X. Shi, and T. Hanamura. 2012. Sustainability issues for underground spaces in urban areas. Proceedings of ICE - Urban Design and Planning [online]. Available: http://www.icevirtuallibrary.com/content/article/10.1680/udap.10.00020 (accessed July 10, 2012).

Thompson, E. 2010. Geohazards Database Consortium [online]. Available: http://gdc.cee.tufts.edu/ (accessed November 13, 2012).

U.S. Census Bureau. 2008. National Population Projection. U.S. Census Bureau [online]. Available: http://www.census.gov/population/www/projections/2008projections.html#. (accessed November 11, 2010).

U.S. Census Bureau. 2012. 1900 Fast Facts. U.S. Census Bureau [online]. Available: http://www.census.gov/history/www/through_the_decades/fast_facts/1900_fast_facts.html (accessed May 4, 2012).

USNCTT (U.S. National Committee on Tunneling Technology). 1989. Micro- and small-diameter tunneling: The U.S. perspective. Tunnelling and Underground Space Technology. 4(2):193-199.

van Outeren, E. 2009. Subway Headaches: Amsterdam Metro Line Sinks Deeper into Trouble. Spiegel, April 7, 2009 [online]. Available:http://www.spiegel.de/international/europe/0,1518,617894,00.html (accessed April 29, 2011).

Wallis, S. 2009. Köln—Speculation and Anger in Aftermath. Tunnel Talk, March 2009 [online]. Available: http://www.tunneltalk.com/Cologne-collapse-Mar09-Deadly-collapse-in-Cologne.php (accessed April 29, 2011).

Ward, J. 2010. Stuttgart 21: A Four Billion Euro Makeover. Spiegel [online]. Available: http://www.spiegel.de/international/germany/0,1518,710388,00.html [accessed April 30, 2012].

Watts, D.J. 2004. The "new" science of networks. Annual Review of Sociology. 30:243-270 [online]. Available: http://www.jstor.org/stable/29737693?seq=1 (accessed November 7, 2012).

Webster, G.S. 1914. Subterranean street planning. Pp. 200-207 in Annals of the American Academy of Political and Social Science: Housing and Town Planning. E.R. Johnson, C.L. King, R.C. McCrea, J.C. Ballagh, T. Conway Jr., G.G Huebner, S.S. Huebner, C. Kelsey, J.P. Lichtenberger, L.S. Rowe, and T.W. Van Metre, eds. Philadelphia: American Academy of Political and Social Sciences.

Wood, A.M. 2000. Tunneling: Management by Design. London: E & FN Spon

WSSC (Washington Suburban Sanitary Commission). 2009. Forensic Analysis Links River Road Water Main Rupture to Improper Installation 44 Years Ago. WSSC Media Release: May 20, 2009 [online]. Available: http://www.wsscwater.com/home/jsp/misc/genericNews.faces?pgurl=/Communication/NewsRelease/2009/2009-05-20.html (accessed November 11, 2010).

3

Contributions of Underground Engineering to Sustainable and Resilient Urban Development

The first two chapters of this report discuss the general attributes of underground space. This chapter examines how underground space use underpins the long-term sustainability of urban areas, what additional research may be necessary to enhance underground engineering practices, and what developments in underground engineering would further support urban sustainability. This report does not develop arguments for specific sustainable urban development approaches; rather, it examines how the underground can support or contribute to those approaches shown or suggested to be sustainable and how underground use directly affects identified sustainability issues. Some key aspects regarding sustainability of urban communities will be briefly explored.

This chapter discusses the urban setting as a system of systems, and the broadest relationships between underground space use and the essential elements for urban sustainability. Physical qualities of infrastructure related to transportation, shelter, food, water, and key material resources that contribute to sustainability or make them vulnerable to hazards are described. The chapter then focuses on more direct relationships in terms of maintaining enduring, livable communities and enhancing risk mitigation through the use of appropriately planned and designed underground facilities. Chapters 4, 5, and 6 examine advances in human safety issues, analytical techniques for lifecycle cost assessment of underground facilities and the broader "triple bottom line" analysis (financial, economic, and social performance), and specific technological advances associated with enhanced sustainability, respectively.

THE BROAD VIEW: THE URBAN SETTING AS A SYSTEM OF SYSTEMS

Sustainability is dependent on more than having enough clean water, food, and material goods. As urban areas grow, strategic growth of infrastructure systems is also necessary to allow for efficient and sustainable delivery of water and sewerage service, food, energy, industrial and commercial goods, and information. Locally created products or services need to be transported or exported, other goods need to be imported, and wastes need to be removed. Physical infrastructure systems are thus critical to the urban system of systems and underpin both a sustainable economy and quality of life.

How does the growth of urban populations, the expansion of urban lands, and their associated facilities and infrastructure enhance or hinder the provision of essential materials and services and the creation of stable, sustainable, socially desirable urban communities? What is the role of the underground? As described in Chapters 1 and 2, the underground is best thought of as a resource designed and managed using a system of systems approach to achieve the most sustainable solutions. Infrastructure is a substantial shaping force in urban and regional development. In developed areas, underground infrastructure may offer one of the few acceptable ways to encourage or support the redirection of urban development into more sustainable patterns because new support infrastructure can be added relatively unobtrusively. A well-maintained, resilient, and adequately performing underground infrastructure is essential to future sustainability of cities. Much, however, can be done to improve the sustainability aspects of underground facilities themselves.

Urban sustainability will be more likely if it becomes the expectation among urban planners and managers that the urban setting includes the space resources both above- and belowground, and that both contribute to the healthy functioning of a city. This chapter discusses some urban resources and their potential roles in a holistic accounting of urban systems; the following section specifically highlights certain uses of the urban underground that greatly contribute to urban sustainability.

Utilidors

Sustainability planning requires forethought regarding operation and maintenance issues for the entire life cycle of the infrastructure. Allowing ease of access for maintenance, repairs, and upgrades is a means of insuring that such work can be completed at lower costs. Experience from subway construction and other large underground works has led to interest among some subsurface utility providers in combining utility services in common utility tunnels—often termed "utilidors" (or "galleries" in Europe; see Box 1.4, Figure 2 for an example of a utilidor) (APWA, 1971). Utilidors provide continuous maintenance access to utilities without the need for digging in the street, are designed to minimize subsurface displacements and other influences that may cause damage to buried and aboveground facilities, and are a more efficient use of underground space than are separately buried utilities. A study by researchers in Spain (Riera and Pascal, 1992) found a distinct economic benefit from locating services in a common tunnel when the value of the underground was included in the calculations during construction of the Barcelona Ring Road. In fact, shared utility tunnels are frequently constructed in Europe where narrow rights-of-way and strong centralized decision making have favored their use.

It has proven difficult to develop utilidors as extensively in the United States. Obstacles include the need to abandon investment in existing service infrastructure, concerns about operational liabilities and risk in a shared or co-located utility environment (e.g., water or gas lines in the same tunnel as electric lines), and administrative concerns related to access to utility lines by others. In addition, initial connection costs may be higher than those for dig and place utilities. Operational issues such as risk and security concerns for utilities, if installed in utilidors, could be circumvented with improved sensor and security systems. The viability, value, and benefits of utilidors may be effectively communicated with (1) development of workable scenarios for secure multi-utility facilities; (2) development of workable scenarios for effective transitioning from current configurations; (3) lifecycle cost-benefit analyses comparing separate and combined utility corridors; and (4) demonstration projects. In the United States, utilidors have been built typically as part of major old and new developments or underground transportation improvements (e.g., Disney World in Orlando, Florida, with its extensive underground service "city" and the Chicago freight tunnel network). If the United States is to improve the sustainability of its urban utility services and preserve underground space for more cost-effective sustainability opportunities for future services, then this impasse needs renewed attention.

Underground Transportation Facilities

The long-term sustainability of urban areas is positively affected by the availability of underground transportation systems. Cities such as Singapore have benefited from master plans designed around transportation systems (Hulme and Zhao, 1999). Well-planned underground transportation systems tend to reduce urban sprawl, saving landscapes and protecting biodiversity, and can positively impact land use and development decisions (Bobylev, 2009; Sterling et al., 2012). They provide safe and efficient transportation and decrease the need for and use of automobiles, reducing congestion and travel times, which in turn reduces fossil fuel use and emissions (Besner, 2002).

Underground transportation assets can address multiple growth-related challenges in urban areas, but many challenges also remain to be addressed (see Box 3.1). Today, many cities have urban transit subway systems, underground

BOX 3.1

Specific Challenges and Opportunities for Transportation Systems

Underground transportation systems will benefit strongly from technical advances as discussed throughout this report. In design and construction, for example, new lining and underground construction technologies are needed that reduce material use and improve long-term facility performance. Underground transportation systems in major cities, however, usually represent key infrastructure elements that are pivotal in terms of the urban mobility that sustains the economy and provides quality of life and hence have a special importance in terms of underground space use. Because they are large public investments and subject to many policy and funding constraints, underground transportation systems may not be designed, operated, and maintained for their maximum contribution to overall urban sustainability. The construction of major underground transportation projects often requires significant relocation of in-situ underground utilities along public rights of way. However, the major excavation work and relocation needs of the project provide key opportunities for renewing and rationalizing utility provision in an area to provide for easier future maintenance of those systems. While this represents an extra burden on the transportation project, it can provide an overall benefit to the urban community using a system-of-systems analysis rather than a project-by-project analysis. Furthermore, in a planning context example, the long-term sustainability of an underground transportation system is improved when system designs allow as much flexibility as possible, taking into account future uses, potential for additional transportation lines, and intermodal connections. This again can increase initial costs but provide for better long-term sustainability.

express arterials and highways, and grade-separated dedicated freight movement corridors for railroads or trucks. High Speed Rail (HSR) service that includes both above- and belowground components is common in Europe and Asia. Each system has unique characteristics to suit its purpose and location. All will likely improve quality of life and long-term sustainability benefits to the urban center(s) served (Jehanno et al., 2011).

Underground transportation, as described in the next sections, can serve to increase community resilience against many natural or manmade hazards including earthquakes and acts of war than their surface counterparts. Box 3.2 provides an example of the performance of transportation infrastructure crossing San Francisco Bay following the Loma Prieta earthquake in 1989. Different types of underground transportation elements and systems and their roles in sustainable urban development are described.

Underground Urban Roads and Highways

Overloaded and congested urban surface arterial roads can be relocated to aerial or underground alignments to obtain grade separation (e.g., transportation routes at multiple elevations) and exclusive rights-of-way. This can relieve the surface of crowded traffic, noise, air pollution, and congestion. The multiple transportation levels provided by tunnels may allow dysfunctional arterial roads to be replaced with functional surface roads that improve the quality of life for neighborhood residents and transportation mobility for the city. The physical barrier and visual blight that an elevated arterial road may represent can be removed. Adjacent neighborhoods once separated by the road may be able to reunite as a community (see, for example, Einstein, 2004). Removing traffic to a tunnel may also result in a brighter and quieter environment, new land use opportunities, and improved neighborhood property values—all indicators of more livable and sustainable neighborhoods (Parker, 2004).

Underground urban roads and highways typically traverse deep below a city from portals at each end that tie into existing service road networks. By going deep, the tunnels avoid building foundations and other in-place services, and leave space closer to the surface for future installations. In most cases tunnels constructed at depth will be the lowest cost among alternative underground solutions if a lifecycle cost analysis is prepared (Parker and Reilly, 2009) and geologic conditions are respected. Barriers to free-flowing traffic can be bypassed, travel times shortened, and carbon emissions reduced for the same distances traveled by surface road. Further, diversion of traffic from streets allows more pedestrian-friendly environments in the city. However, decisions to build underground roadways, regardless of the benefits, are regularly contested (for example in Seattle, Washington; see Box 3.3). The decision to proceed often requires a vote of the people and a coming together of city, county, state, and federal representatives to reach agreement. This process is often time consuming and can result in increased project costs.

Public Transit Subways

Public transit is a vital part of many urban areas and an integral part of a sustainable urban environment. Rapid transit facilitates efficient movement of people of every economic class and ethnic group to and from their homes, school, work, health services, places of worship, airports, recreational activities, and other amenities available to urban life. Public transit provides needed mobility to those without cars, and connects and unites neighborhoods and communities to function more smoothly and take advantage of community services. Many cities make public transportation available in the form of bus systems. As populations grow to between 1 million and 3 million, regions may see advantages in electrified rail transits (light rail) (APTA, 2009) that allow faster transit for larger numbers of people. Such systems can operate on streets used by normal traffic, in limited access rights-of-way, and exclusive and grade-separated rights-of-way (for example, elevated or underground as developed for the Muni transit system in San Francisco, California, and the MAX transit system in Portland, Oregon). Heavy-volume transit systems—so called "heavy rail" systems—are needed when populations increase to more than 3 million (APTA, 2009). These are grade

BOX 3.2

Performance of Transportation Infrastructure Following the Loma Prieta Earthquake, 1989

Underground transportation systems can remain operational during, or quickly resume operation following, natural hazardous events such as earthquakes, tornadoes, lightning, and thick fog or dust conditions. According to a review of several studies documenting earthquake damage, large diameter underground tunnels have historically suffered less damage than surface structures (Hashash et al., 2001). The San Francisco Bay Area Rapid Transit (BART) system operates through cut-and-cover and mined tunnels and serves multiple destinations including San Francisco and Oakland, California, through a 5.5 kilometer subaqueous trans-bay immersed tube tunnel between the two cities. This system improved disaster resilience for this urban area following the Loma Prieta earthquake in 1989 by allowing the continued functioning of the economies of these communities.

The Loma Prieta earthquake was a magnitude 6.9 event that caused serious physical damage to local infrastructure (USGS, 2009) including damage to connections, bearings, and members of the San Francisco-Oakland Bay Bridge, forcing its closure for more than a month. A 15 meter, 5-lane roadway section dropped from the upper eastbound roadway deck onto the lower westbound deck (see the Figure), killing one person (Dames and Moore's Earthquake Engineering Group, 2004).[a] BART crosses San Francisco Bay underground almost directly beneath the Bay Bridge alignment. It was temporarily shut down by the earthquake, but there were no passenger injuries, and service resumed in half a day following damage inspection and power restoration. BART patronage rose quickly from an average of 218,000 riders per day to more than 308,000, and service continued around the clock, seven days a week until the Bay Bridge reopened more than a month later (Dames and Moore's Earthquake Engineering Group, 2004). The Bay Area economy,

separated, often in subways such as in the BART system constructed in 1962 in the San Francisco Bay area, and the New York City Transit System, constructed beginning in 1900 (Bobrick, 1981).

Subway rapid transit provides the same safe, environmentally sound, fast, low-cost, and comfortable transportation to all people who use it. It has already been mentioned that choosing subway transit because of its relative comfort, savings in time and money, or predictability of the ride reduces the number of commuters on surface roads. Commuters who use rapid transit daily rather than drive personal vehicles cut their carbon footprint significantly (APTA, 2008), and may realize personal health benefits through minimizing stress associated with traffic, accidents, and congestion. From the regional perspective, regional transit system stations attract development of urban centers—small urban communities—because access to the urban areas becomes a major attraction for those relocating to the region. The location and services that support more dense, compact development in the vicinity of transit stations—as opposed to the development of urban sprawl—can affect the overall cost to the taxpayers in terms of provision

although damaged by the earthquake, recovered more quickly than would have been the case without the underground BART because significantly large numbers of people were able to get to work (USGS, 1998).

FIGURE Collapsed section of the San Francisco-Oakland Bay Bridge following the 1989 Loma Prieta Earthquake. The bridge remained closed for more than a month while the BART subway tunnel located almost directly beneath the bridge was running within a day of the earthquake. SOURCE: sanbeiji (CC-BY-SA 2.0), available at http://www.flickr.com/photos/sanbeiji/220645446/sizes/m/in/photostream/.

[a]Similarly, the upper roadway of a 2 kilometer length of highway of the Cypress Street Viaduct in the San Francisco Bay area crashed onto the lower roadway, killing 42 and injuring several hundred more.

of essential services such as schools, police, fire and EMS protection, hospitals, water, sewer, electrical, natural gas, food, and other supply sources, all necessary attributes of developing a sustainable urban environment.

The ability to update and replace subway system components such as conduit, electrical and fiber optic cables, water lines, waste water lines, ventilation systems components, lighting, signage, escalators and elevators, and information systems makes it reasonable to expect useful service of subway tunnels for more than 100 years. Transit tunnels built in the 1860s in London are still in service today. The long life of underground components tends to reduce lifecycle costs and also reduce demands for both renewable and non-renewable resources (Parker, 2004). All these characteristics contribute to sustainability and justify new rapid rail subways from a lifecycle analysis point of view.

Grade Separated and Underground Freight Railroads

Combining normal surface and freight traffic, particularly the movement of ubiquitous freight container units, can result in heavy traffic, especially in port

BOX 3.3

Replacement of the Alaskan Way Viaduct, Seattle, Washington

Recent experience in Seattle, Washington, planning the replacement of the earthquake-damaged Alaskan Way Viaduct (AWV) illustrates how difficult the decision to reroute to the underground can be. The current AWV is a double-deck urban expressway (see Figures 1 and 2) running along the Seattle waterfront. It is similar in design and construction to the 1950s-era San Francisco Bay Area Cypress Street Viaduct and Embarcadero Freeway that both failed as a result of the Loma Prieta earthquake in 1989 (USGS, 2009). The AWV sustained non-reparable damaged as a result of the 2001 Nisqually earthquake (PNSN, 2002) and must now be replaced (WSDOT, 2004). Alternative solutions included a new, wider, two-level viaduct on the same alignment, a replacement of the viaduct by a wide surface street carrying significant levels of through traffic, the relocation of the highway on a bridge or tunnel over or under Elliott Bay, and the “do nothing” alternative intended to limit traffic growth and create a demand for better public transit through continued, more disruptive road congestion. The alternatives were studied and publicly discussed. Ballot measures to determine the preferred solution were intensely debated at the local, city, and state levels.

Ultimately, the decision was made to bore an urban underground bypass expressway, remove the damaged viaduct, and restore an accessible scenic waterfront (see Figure 2). The 3.2 kilometer, four-lane bypass roadway tunnel (Figure 3) will be located deep enough under the city to avoid the century old Burlington Northern Santa Fe railroad tunnel in daily use, a large interceptor sewer, and existing building foundations. Seattle will recover views of Elliott Bay, Puget Sound, and the mountains when

FIGURE 1 Seattle Washington’s Alaskan Way Viaduct is a double-deck expressway along the city’s waterfront. SOURCE: http://en.wikipedia.org/wiki/File:Alaskanviaduct.jpg.

the viaduct is removed. Major disruption of traffic patterns to and through downtown Seattle will be avoided (FHWA, 2011). A landscaped boulevard on the waterfront is planned, similar to that constructed in San Francisco following the failure of the Embarcadero Freeway. Negative effects of the old viaduct on the city were not fully appreciated until the debate for its replacement took place (e.g., Garber, 2009; Lindblom and Heffter, 2009). Proponents of the plan argue that downtown Seattle will benefit from improved open spaces and green zones.

FIGURE 2 (Left) Arial view of Seattle, Washington, water front and the prominent Alaskan Way Viaduct and (Right) early concept of proposed new Alaskan Way Street of same area. The new concept increases pedestrian access to the waterfront and improves general access to adjacent commercial enterprises. SOURCE: WSDOT.

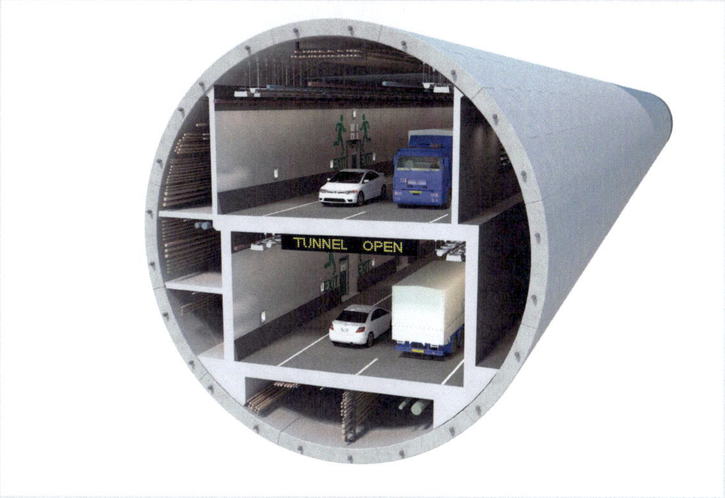

FIGURE 3 Early concept of the proposed State Road 99 bored tunnel. SOURCE: WSDOT.

BOX 3.4

Grade Separation of Freight in Greater Los Angeles

In the greater Los Angeles area, the Ports of Los Angeles and Long Beach have grown to be, if taken together, the largest container terminal in the country (AAPA, 2011). They provide a major gateway for containerized goods in and out of Asia. The principal mode of transport of containers away from the ports to the rest of the country is rail. Three major railroads—the Southern Pacific, the Union Pacific, and the Burlington Northern Santa Fe (BNSE)—had tracks into the ports from their national track junctions. Historically, railroads have right-of-way over crossing traffic, and so much freight movement at grade brought traffic in a large area of southern Los Angeles County to a standstill multiple times daily as 200-car-long freight trains moved slowly over three separate rail networks to join their national track networks east of the urban area.

Concerns over congestion and associated air pollution led to the development of the Alameda Corridor Project (ACTA, 2012a), a plan to build a 32-kilometer (20-mile)-long freight rail expressway including a 16-kilometer (10-mile)-long top braced open trench, 15 meters wide and 10 meters deep with space on its floor for three tracks and a service road called the "Mid-Corridor Trench." The Alameda Corridor Transportation Authority (ACTA) was proposed, organized, and authorized by legislation (ACTA, 2012b). ACTA has authority to raise funds, receive government grants, own and receive property, contract for construction and operations, and do those things necessary to implement the plan. In 1994, with the purchase of the Southern Pacific Railroad's Alameda Corridor track and right-of-way, the corridor project began in earnest.

The 10 meter (33 feet) depth of the Mid-Corridor Trench easily provides the ability of BNSF Railway and Union Pacific Railroad, via their trackage rights, to move double-stacked container freight rail flat cars, 200 at a time, in both directions, at 40 mph from the ports to their respective national rail system connection (ACTA, 2012a). First operations began in 2002, and the more than 200 at-grade railroad crossings where cars and trucks previously had waited

cities. Drivers may encounter long lines of traffic waiting for freight trains to clear grade crossings or trucks in long queues waiting to clear signalized intersections. Significant air pollution from train and truck exhaust, as well as from the traffic waiting to pass, can degrade air quality (Hricko, 2006) and has the potential to negatively impact the quality of life and the economies of nearby neighborhoods (for example, Palaniappan et al., 2006).

Grade-separating freight movement from surface streets is part of the solution. Open braced trenches that provide natural ventilation for diesel exhaust have been a preferred solution in places such as southern California for freight trains powered by diesel-electric prime movers (see Box 3.4). In southern California, significant investment in grade separation infrastructure is the result of collaboration between the ports, a number of affected cities, the county, state, and federal governments, and the railroads.

Some traffic problems can be eased with dedicated and signalized surface

for trains to pass have been replaced with bridges crossing the trench, restoring traffic circulation and keeping local neighborhoods connected.

Outcomes include sustainability benefits for the region, and operational improvements for the ports and railroads, restoring some of their competitive edge by decreasing freight delivery times. Peak movements were reached with 60 train movements per day in October 2006. Benefits to air quality result from more direct rail routes traveled at greater speeds, reduction of vehicular exhausts at grade crossings, and the increase in the amount of cargo that can be transported by rail instead of by truck (Weston Solutions, 2005). ACTA is designing and will soon construct the Alameda Corridor East project, with more braced trench design and a $500 million construction project to grade separate the long freight trains from the grade crossings throughout a part of the city of San Gabriel.

FIGURE A container train of the Alameda Corridor Freight Line in California. The trains travel in an open-braced trench that provides ventilation for the engines and grade separation for container traffic. SOURCE: Courtesy of the Alameda Corridor Transportation Authority

streets or grade-separated viaduct roadways for freight movement by truck during times other than commute periods. Tunnels also can be used to provide exclusive or preferred lanes for freight movement by truck. For example, in Miami, Florida, a tunnel boring machine-driven tunnel beneath Biscayne Bay is being constructed to create a direct connection from the Port of Miami to local highways and reduce traffic in the downtown area (Port of Miami Tunnel, 2010). In the greater New York metropolitan area, tentative planning has begun again on a freight-only tunnel that would pass under a part of Eastern New Jersey, the Hudson River, Manhattan Island, and part of Brooklyn, New York, possibly providing for the movement of freight trains and trucks between the vicinity of the New Jersey Turnpike and the Long Island Expressway (FHWA/PANYNJ, 2010). The ambitious plan indicates growing recognition that not enough surface area exists to provide for the needs and services required to remain competitive in a global market. Grade-separation tunnels would be a part of the answer, and sustainability benefits will be among those analyzed by planners to justify such a facility.

There has been persistent recurring interest in underground pipelines for transporting parcels or freight, for example the use of extensive networks of pneumatic tubes in Paris, France, Vienna, Austria, Berlin, Germany, and Prague (now of the Czech Republic), since the second half of the late 1800s (Uffink and Admiraal, 2012). Such systems involve the transport of freight in a capsule, propelled by liquid (hydraulic capsule pipelines) or gas (pneumatic capsule pipelines) through a network of pipes (Liu, 2000). Pneumatic capsule pipelines were used in the former Soviet Union, and a commercial system is in place in Japan for transport of construction materials, limestone, and similar commodities (Liu, 2000). Other systems have been studied elsewhere in the world, particularly in port areas (Uffink and Admiraal, 2012). Specific plans for U.S. freight tube systems have been studied (e.g., Goff, 2001; Roop et al., 2003; Liu, 2004) but have not been implemented due to cost competition from other modes of freight movement and various environmental issues.

High Speed Rail

Broader sustainability benefits can accrue when regional rail transportation systems compete with airlines in terms of travel times and costs. High speed rail (HSR), for example, could deliver large numbers of people over long distances with a significantly smaller carbon footprint (Baron et al., 2011; Ledbury and Veitch, 2012). HSR systems are in service in Japan, France, Germany, Italy, China, and Taiwan, and accommodate home-to-work trip commuting by providing long-distance point-to-point transport on a reliable schedule. In the United States, California voters passed a bond issue in 2008[1] enabling the construction of HSR from Los Angeles to San Francisco (see Box 3.5).

HSR is typically planned with no more than 1.0 to 1.5 percent vertical grades to maintain maximum speeds, and very long radius vertical and horizontal curves to accommodate high ground speeds of up to 220 mph.[2] Higher grades and smaller turn radii can be accommodated with commonly used technologies, for example grades of 3.5 to 4.0 percent on some HSR lines in Europe, but speeds may be compromised. These speeds require that rights-of-way be exclusive and protected from access to other vehicles, people, or large animals, and to achieve such HSR makes use of viaducts, open top trenches, and tunnels. When HSR approaches a destination city, it slows and slips underground to penetrate the city center below existing infrastructure. One or more large underground rooms are located beneath the city center to house the station, supporting facilities, and personnel required to deliver the service product. Vertical delivery systems lift

[1] CAL. S.B AB 3034 (2008).

[2] See, for example, http://www.cahighspeedrail.ca.gov/project_vision.aspx.

BOX 3.5

High Speed Rail in the United States: The California Example

High speed rail (HSR) may become a reality in the United States in the next two decades. HSR can be economically viable in between major metropolitan areas that are 160-800 kilometers apart, with sufficient populations (50,000 and higher) and economic productivity and a demonstrated travel market (Hagler and Todorovich, 2009). State and federal support, funding, and financing would need to accompany local interest in HSR service. Planning, environmental impact statements and mitigation measures, rights-of-way acquisition, and design and commitment to construction efforts would be significant, and engineering, geotechnical, historical, and archaeological investigations and reports would be important to the timely development of the project. Much earth science and seismic investigative and analysis work would be required to support the design solutions, especially given the need to survive and remain in service following a significant seismic event.

In 2008 California voters passed a $10 billion bond issue to enable the construction of an HSR system between Los Angeles and San Francisco, California, eventually to be extended south to San Diego and north to Sacramento. It is in the early stages of conceptual design, working through alternative designs, section by section, and filing state environmental impact reports. Total track length is estimated to be 2,775 kilometers, with most of the line to be two-track. Between 160 and 200 track kilometers of tunnel are planned. San Francisco and Los Angeles stations will be reached through deep tunnels, delivering passengers to the Transbay Terminal in San Francisco and to Union Station in Los Angeles. The tunnel designs are being developed for train speeds of 350 kilometers per hour (217 miles per hour). To provide the needed rights of way in other populated areas, significant use of open cut, open top, braced reinforced concrete box structure system for two operating tracks and a service road are being considered. The cross section is similar to that used with the Alameda Corridor Project.

passengers and luggage to street level. HSR, however, is very expensive and often subsidized.

Housing

Shelter in urban areas in the United States ranges from low-density developments of single-family homes to high-density apartment and condominium properties. Low-density developments offer self-reliant sustainability possibilities in terms of on-site energy collection, food production, and local management and recycling of some wastes generated by the occupants. However, energy expenditures for transportation in low-density urban areas greatly exceed those in high-density urban areas (Newman and Kenworthy, 1999) because of longer travel distances and limited public transportation options. Providing centralized services to low-density developments requires increased lengths of utility

FIGURE 3.1 The earth-sheltered home is partially underground, providing greater strength and energy efficiency. SOURCE: http://www.monolithic.com.

services as compared to the same population served in high-density areas. High-density housing increases the dependence of occupants on centralized services, but increases options for a non-automobile-based urban lifestyle, because public transportation can be provided more economically and basic shopping opportunities can exist within walking distance for many residents. Urban development trends in the United States show a continued increase in urban sprawl, but also a trend toward increased population densities in urban downtowns (for example, Greene, 2006), motivated in part by the desire for a more urban living experience without long and expensive commutes by car.

Most people do not choose to live underground. Urban underground use related to housing is in the form of utility and transportation services for residents, storage, or expanded living space (e.g., basements). In low-density suburban and rural housing developments, some earth-covered or earth-bermed structures have been built for their ecological, isolation, and energy attributes (see Figure 3.1), but initial costs, moisture control issues, acceptance, and resale issues limit their widespread construction.

Urban Commercial, Industrial, and Institutional Facilities

Sustainable urban areas must provide necessary commercial, industrial, and institutional infrastructures that support and manage a viable economy, provide jobs, and deliver support including education and social services. The relationship between urban density and land prices tends to create a market for large, multi-story, commercial and institutional buildings in the high-density core(s) of urban areas, and more low-rise structures in the surrounding urban and suburban areas. As discussed elsewhere in this report, it can be desirable under certain circumstances to place various types of commercial, industrial, and institutional facilities underground. However, this is not the only use of the underground that is important in the urban core. Increased size and density of buildings in the urban core require an increase in the capacity of urban utility and transportation services, both in the core and in outlying areas that are connected socially and economically. Increased reliance on such systems means that the systems must be robust and reliable. The underground can offer protection to these systems, keeps them near the populations they serve, and allows provision of critical lifeline services and emergency response—all while preserving aboveground real estate for other uses.

Sustainable Food Production and Distribution

Prime agricultural land is being covered by low-density suburban development in many parts of the United States (Carver and Yahner, 1997). There has been a historic tendency for populations to concentrate in areas with good agricultural, trading, or transportation potential, often along rivers or coastlines. Hence, as cities expand, they often sprawl out from the population center and supplant good farmland. Many existing infrastructure and taxation practices actually encourage urban sprawl by supporting construction of new regional infrastructure needed to service the growing suburban areas (Brueckner, 1999). The market forces that underlie conversion of farmland to developed land represent important long-term issues that have potential impacts on the cost, availability, and impact of food supplies. Abandoning good farmland close to markets and developing poorer farmland farther away must be balanced by efficiency improvements if regional sustainability is to be achieved.

Placing facilities underground reduces claims on surface land, which can have the effect of preserving agricultural land. Careful holistic urban planning and placement of underground facilities can serve to direct urban growth in the most sustainable manner. Holistic planning considers not only how productive agricultural lands may be preserved, but also how food is made readily accessible to urban populations through transportation infrastructure, temperature-controlled storage facilities, means of distribution and sales, and the energy necessary to operate each part of the delivery/storage chain. Maximum sustainability benefits are achieved if all of these infrastructure requirements are considered as part of the overall urban system of systems. Such facilities do not necessarily need to be underground, but underground facilities do provide some inherent thermal advantages in terms of food storage and warehousing. Warehouses and retail food outlets, for example, are typically windowless facilities that have little need to occupy the surface, especially when surface space is scarce.

Water Resource Preservation and Distribution

Water is essential to human survival, and loss of a hygienic water supply impacts survival and can cause the spread of disease more rapidly than loss of food supply. Urban areas typically draw their water from surface water (rivers, lakes, and reservoirs) or groundwater sources. Water resources are controlled largely by the hydrologic cycle, but can be damaged by poor practices associated with land development. As urban areas grow and reliable and hygienic surface and groundwater supplies are no longer adequate to meet demand, efficient water use, reuse of "grey" water (wastewater from domestic activities such as bathing and laundry) and "black" water (sewage), and creation of new water supplies become more important. Cities, especially those in arid regions, are forced to seek ever-more-distant or deeper water supplies as urban populations grow and the demand for water increases. Aging or poorly maintained water supply infrastructure leaves urban water supplies vulnerable to leakage[3] and to disruption. There is also growing competition for water between urban areas and agriculture regions where irrigation occurs on a widespread scale (FAO, 2011). Holistically thinking, the export of agricultural products from a region can be considered a loss in water resources—even as it may be a boon to the regional economy.

Long-term urban sustainability implies that an urban area has balanced its water supply possibilities. Where groundwater is an important water resource, a sustainable water supply would not be depleted (for example, by over extraction), polluted, or diverted in detrimental ways (see Box 3.6). Construction activities may produce runoff with sediments and pollutants, such as pathogens and metals, that may negatively impact water quality or quantity (EPA, 2005). Agricultural practices may also have long-term impacts on water quality and the regional environment. For example, increased salinity of surface water and groundwater caused by evaporation and dissolution (CA Water Resources Control Board, 2010), and attendant changes in habitat for flora and fauna, can be the direct result of water use practices.

Good stewardship of underground water resources can include use of the underground for urban development because more natural landscape can be preserved for groundwater recharge. Water distribution facilities can be placed underground, allowing land to be developed in ways that enhance quality of life in compact cities. However, careful analysis and construction and operation approaches are needed to avoid detrimental changes in groundwater levels, flow patterns, and pollution. Groundwater pollution can become a major issue; the pollution legacy of underground gasoline storage tanks at service stations is a well-recognized example (Meehan, 1993), as are superfund cleanups of major industrial pollution sites.[4]

---

[3]For example, see Hull, 2010.

[4]See http://www.epa.gov/superfund/.

BOX 3.6

Groundwater Flow Under the Duisburg (Germany) Subway

Underground tram stations in Duisburg, Germany, were designed for future traffic expansion with two platform levels and cross-platform changing in each direction. In 2000, a 3.6 kilometer extension of the Stadtbahn tunnel from north under the river Ruhr to Meiderich opened (UrbanRailNet, 2007). When the Duisburg line was being built, the slurry (diaphragm) walls were carried down to a clay layer aquaclude (impermeable layer) and would have permanently impeded northward flow of water under the city, causing a buildup of water levels on the upstream (southern) side and dropping water levels on the downstream (northern) side. This potential change in groundwater conditions was unacceptable. The Municipality of Duisburg and the prime contractor designed a system that would be watertight during construction then permeable after construction. Dense bentonite (clay) slurry would fill and support the sides of the trench until cast-in-place concrete or concrete panels replaced it. An approximately 1.3 meter gap was built between 5.4-meter-long cast-in-place panels of impermeable diaphragm wall. The gaps were frozen before excavation for the necessary ground support, creating an impermeable barrier. The freezing pipes were removed when construction was finished and the slurry thawed, permitting groundwater to pass under the tunnel (Hooks et al., 1980).

In this circumstance, issues beyond construction were considered, and a solution to a potentially serious problem was applied. However, this example illustrates the potentially large-scale problems that can occur during subsurface construction without the appropriate sensitivity to the impacts of underground design, construction, and operation on an entire urban system.

Key Material Resources

Sustainable use of a non-renewable resource seems contradictory, but in practical terms, sustainable use can be considered a question of the rate of use of a resource over a meaningful timescale. Key material resources derived from earth materials, for this discussion, fall into the categories of fluid and gaseous energy sources (principally oil and gas), energy sources in solid form (e.g., coal, oil shale, tar sands, wood, and peat), industrial minerals (e.g., iron ore and bauxite), high-value or critical strategic minerals (e.g., gold, uranium, rare earth elements), and construction-use materials (e.g., gravel, sand, building stone, Portland cements, and brick-quality clay). Consideration of the interdependencies of the energy and mineral sources, available reserves, strategic concerns, and environmental impacts is beyond the scope of this document. However, issues that arise in the interaction between urban development, energy and mineral extraction, and the use of urban underground space are justified for discussion in this report, and brief and general descriptions of some issues are provided.

Estimates vary for how long world oil, gas, and other energy reserves will last given current consumption rates. Usage rates and the extents of proven accessibility of reserves depend in part on the unit prices of the resources.

Oil and gas are important feedstock for the plastics and chemical industries, complicating price and policy interactions. When alternatives are available, there may be stronger incentives to switch to alternate resources. Wind energy, solar energy (e.g., thermal power, passive solar heating, and photovoltaic electric energy), biomass (e.g., rapidly replenished sources for hydrocarbon fuels and direct burning for electric power generation), geothermal (both hot rock power applications and ground-coupled heat exchange systems), and wave energy capture systems are among the more commonly discussed options. Considering these options in the long term is essential in urban development, including making the means available to respond to growing energy demand as energy sources and technologies change.

Industrial minerals include a range of ore types and extraction methods. Most markets are price sensitive, and many industrial ores are mined outside the United States. Such globalization has resulted in the closing of many U.S. mines and, consequently, fewer options for educating mining engineers in the United States (see Chapter 7 for additional discussion of this topic). Expending greater resources to extract minerals of high strategic importance may be justified, but domestic sources are often ignored in favor of less expensive foreign sources (NRC, 2008a). Construction materials are mostly constrained to local extraction, transport, and use because of their high bulk and low price potential. They are, however, important to urban building and infrastructure construction. Poor local availability can significantly increase construction costs and hamper development.

Important natural resources may gradually become inaccessible as a result of urban development—essentially becoming quarantined beneath expanding urban areas. New technologies may be needed for successful and safe extraction in developed areas. For example, local aggregate resources are frequently obtained from nearby open gravel pits or open rock quarries. As an urban area encroaches on the resource, open excavations become an increasing nuisance of noise, dust, vibrations, competition for road transportation capacity, and other disturbances. Land value increases also may encourage sale of the land for development.

An alternative to abandoning rock quarry resources is a change to resource recovery through underground mining. Under the right combination of rock quality and economic conditions, an aggregate supply can be maintained and newly created underground space in large mined caverns can provide a stable natural temperature and a high degree of separation from other urban or recreational uses on the land surface above. The Kansas City area is a prime example of what can be developed. Approximately two-thirds of the industrial space in the Kansas City area is located in large mined limestone caverns (Nadis, 2010) (see Box 3.7). Other mined spaces around the world have been used as facilities to store everything from paper and electronic archives; energy, waste, and agricultural products; frozen foods; and compressed air. They also have been used as museums and tourism facilities, sports facilities, education facilities, hospitals,

BOX 3.7

Underground Commerce Centers: Turning Mining Excavations into New Commercial Resources

Developers in U.S. locales such as Lawrence County, Pennsylvania, and Kansas City, Missouri, have converted their underground excavations into warehouse, office, manufacturing, and educational space (see Figure). Limestone mining began in Kansas City in the late 1800s. By the mid-1900s, mine owners strategically excavated to utilize space left behind (Buzbee, 2011). "SubTropolis" encompasses approximately 7.62 million square meters of leasable space. As of 2010, 55 businesses were located in the underground facility that includes a 3.2 kilometer network of rail lines and 9.7 kilometers of paved roads (Nadis, 2010). Constant underground temperatures result in 50 to 70 percent savings in total energy costs (Hunt Midwest, 2009). No heating costs are incurred in the winter, and very little energy is required for cooling and humidity control in the summer. Space use is diversifying as the company develops a 61,000 square meter data center with redundant power and cooling systems and protection from natural disasters.

In 1991, the facility had a major and difficult-to-control fire that burned for weeks in a large storage area in spite of firefighter efforts to control it. The underground fire was too hazardous for direct fire suppression by fire fighters, and no fixed fire suppression system such as sprinklers were in place at the time. Cleaning compounds, pesticides, paper goods, and cooking oil contributed to the fire that approached 1,100°C (2,000°F) (Buzbee, 2001). Similar problems were encountered at another underground facility located in Louisville, Kentucky. As a result of such fires, the National Fire Protection Association established a Technical Committee on Subterranean Spaces, and new fire protection standards were developed related to distance to and numbers of exits, ventilation, communication in the underground, and underground wayfinding (Lake, 1998). Kansas City has since adopted new safety language into its code for underground spaces that establish minimum safety requirements.[a] SubTropolis and other underground facilities now include fire suppression systems and safety practices.

FIGURE Underground warehouse space in mined limestone caverns approximately 100 feet beneath Kansas. Roads and facilities can accommodate 18-wheeler traffic. SOURCE: Hunt Midwest, http://huntmidwest.com/press.html.

[a]See http://ww4.kcmo.org/codes/CH18/2006/CH18ART11.pdf (accessed June 9, 2011).

laboratories, and for a variety of retail, office, and manufacturing purposes (Peila and Pelizza, 1995). Of course, such mining must be executed carefully to ensure the long-term stability of the surface and underground space. There are many examples around the world of cities growing over old mine workings that were not excavated with long-term stability in mind. Such mines may pose collapse and subsidence hazards for surface development, as well as provide a pathway for degradation of area groundwater through leaching of harmful chemicals from the mine workings.

Local Urban and Natural Environment Preservation

Beyond having food, water, shelter, a viable economy, and a supply of key resources, a sustainable urban environment requires that natural processes are sufficiently maintained to preserve an ecological balance over the long term. A sustainable and healthy urban environment is fundamentally linked to a sustainable and healthy natural environment. A deteriorating natural environment may directly and deleteriously impact food and water supplies, and ultimately degrade quality of life and health to unacceptable levels. Key environmental parameters for sustained urban quality of life include carefully considered air and water quality standards, noise control, and safety and sanitary standards. Basic living standards in even the poorest neighborhoods need to be met for urban systems as a whole to be sustainable. Aspiring beyond basic levels of environmental preservation means creating an urban environment that is appreciated by all citizens and that offers a variety of social, cultural, and recreational opportunities with easy access to the natural environment. As has been discussed, underground facilities can have many specific impacts on preservation of the surface environment at the site of a facility, or along a transportation network. In broad terms, placing facilities underground allows preservation of more natural space for the benefit of the community.

Underground construction—as compared to surface or elevated construction—can mitigate noise and vibration and can offer greater air quality control and beneficial reuse of waste construction materials, including soil and rock removed from the site. On the other hand, to contribute to the sustainable urban-natural environment system, infrastructure placed underground needs to be constructed with consideration of issues associated with water quality, groundwater flow, potential changes to soil geochemistry, or changes in underground temperatures or heat flow that might impact the natural and built environments.

To limit future pollution, current groundwater, soil, and infrastructure monitoring practices for urban areas may need to be intensified to more effectively identify and address potential problems (see Chapter 6 for detailed discussion). Sustainable practices suggest that environmental problems need to be looked at comprehensively and on a common risk-cost-reward basis.

HAZARDS, SECURITY, AND RESILIENCE OF URBAN AREAS

The form and operation of urban areas has regularly responded to known risks. In historic times, threat of attacks resulted in establishment of walled cities and secure water and food supplies that could last many months. Massive fires, such as the Great Fire of London in 1666, led to changes in both construction practices and street design in major cities (Schofield, 2011); and attention to improved water supply and sanitation in the modern era occurred when the link between cleanliness and disease was established. Natural underground conditions or phenomena such as presence or absence of gases, radiation (radon), excessively high or low temperatures, and water may represent hazards to underground infrastructure and people in or dependent on it. For example, gases such as methane, sulfur, and carbon dioxide naturally exist underground and can threaten human health in certain concentrations or exposures. The lack of naturally occurring oxygen is also a hazard. Human habitation of the underground, therefore, requires continuous ventilation from the surface in a failsafe delivery system. Water also poses a hazard to underground infrastructure and its occupants and can swiftly inundate and damage subterranean structures and safety systems (see Box 3.8). Such risks can be minimized and managed at acceptable levels, but only if identified, understood, and responded to. A successful risk management strategy is one that is tightly integrated with design and operations processes. Robust monitoring systems are needed that ensure overall performance, and human and technological capacities are needed that can design, operate, and respond to substandard performance when encountered.

In recent years, different or new types of hazards have been recognized in the urban environment and require increased attention. Some are related to ongoing urban concerns such as air quality, personal safety, and security; others are associated with vulnerable and deteriorating urban infrastructure systems. Another type of hazard is linked to extreme events including those associated with war, terrorist acts, and natural disasters. Existing data resources are too sparse to allow thorough understanding of complex systems' responses to extreme events or allow reliable behavior modeling and prediction. Extreme events can represent opportunities for large-scale demonstrations, responses can be observed, and design and performance prediction through computational simulations can be improved. This requires pre-organization and preparation (including identification of funding) as well as identification of the cross-sector teams that can be rapidly mobilized to investigate in the aftermath of an event. Teams are mobilized to investigate the aftermath of major earthquakes worldwide, but the focus proposed here is on the understanding and validation of interdependency models.

Security needs also have changed markedly from even 10 years ago, and planners and engineers need to mitigate hazards and risks not previously examined while maintaining societal expectations for well-being and quality of life. Further, facility, materials, and space usage have changed over expected infrastructure

BOX 3.8

The Great Chicago Flood of 1992

"The Great Chicago Flood of 1992" occurred the morning of April 13, 1992, as a result of the placement of a support pillar into the Chicago River bottom during construction work. The ceiling of an antiquated tunnel located beneath the river was damaged, and extensive flooding jeopardized human lives and severely damaged the infrastructure of the Chicago business district (Arnold, 1992). The tunnel is part of a system (see Figure) that ranges from depths of 6 to 15 meters below the river (Inouye and Jacobazzi, 1992); 946 million liters (250 million gallons) of water flooded downtown sub-basements (cbs2chicago.com, 2007). After unsuccessful attempts by city workers and contractors to plug the hole, Mayor Daley asked President Bush to declare the Chicago Loop a national disaster area. On April 15, the Federal Emergency Management Agency was charged with the federal response to the disaster, and the Corps of Engineers began its work with the contractors to carry out the plugging operation. The work was completed 37 days later (Inouye and Jacobazzi, 1992).

Intended to carry telephone and telegraph wires and cables, the 100 kilometer hand-constructed freight tunnel system constructed in the early 1900s was used to transport merchandise and remove solid waste from more than 80 buildings until no longer viable. The tunnels are now used to store power and fiber-optic cables (Wren, 2007). The river flooded the tunnel network "crisscrossing downtown Chicago and connecting to building basements" (Wren, 2007, p. 35). Researchers replicated the tunnel failure in a geotechnical model and explained the dramatic load increases, breach, and flooding. The southeast abutment of a bridge was previously protected by two dolphin pile clusters (tight clusters of piles). During renovations, the piles were removed and the breach was caused by the driving of new piles 1 meter to the south—and closer to the tunnel. The breach was discovered before flooding started, and repairs to the tunnel were planned. Flooding was delayed by slower water seepage through relatively impermeable soil, but the soil ultimately became displaced when the piles were removed. A conduit formed between the river bottom and the tunnel (Wren, 2007).

Reports vary about the economic and human costs of the extensive flooding. The total contract cost for "dewatering" and structural repairs was reported to be approximately $5.5 million (Inouye and Jacobazzi, 1992). The flooding shut down the Loop, a major financial and retail center and seat of government for Chicago, at an estimated cost of between $1 billion (Wren, 2007) and $1.95

life cycles. Few codes and regulations have been developed for application specifically to underground facilities, and fewer still accommodate changing security needs. Wisdom is required in the choice of new codes, regulations, and metrics to measure success. Sustainability is dependent on the ability of planners and engineers to anticipate and be flexible to emerging issues, technologies, and societal expectations over the duration and beyond the life cycle of the infrastructure they design, build, and operate. They need to accommodate the constantly evolving urban environment.

billion (cbs2chicago.com, 2007). It was reported that the Chicago Mercantile Exchange lost $25 billion in trading. Thousands were affected because people had to be evacuated, subways were shut down, buildings and businesses were left without power (some for several days), and the flooding destroyed everything from merchandise and restaurant food supplies to government records. The lower floors of the Art Institute building also were damaged. Eight city officials, including the acting commissioner of the Department of Transportation, either resigned or retired, were held accountable because they knew in advance that a potentially serious problem had developed at the breach site (cbs2chicago.com, 2007).

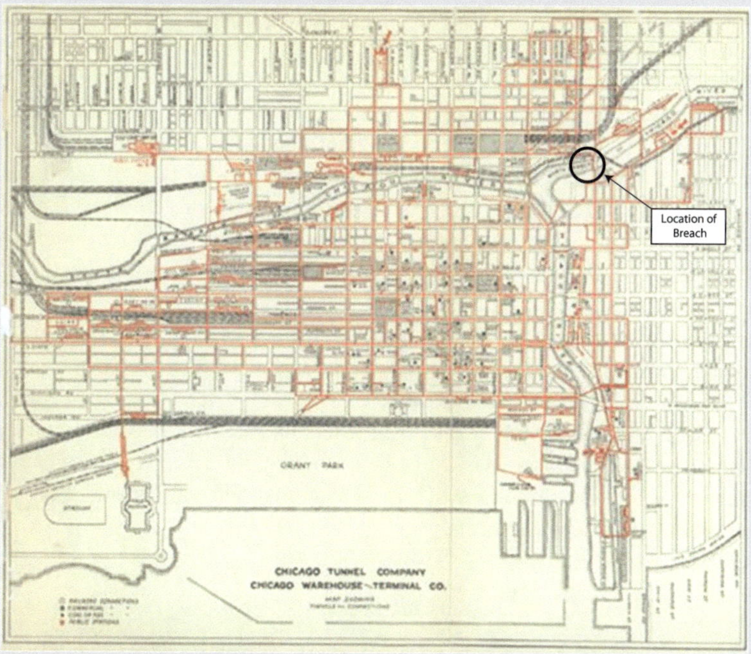

Figure Map showing the Chicago freight tunnel network in 1928 (courtesy of the Chicago Department of Transportation) and the location of the 1992 breach that flooded much of downtown Chicago. SOURCE: CDOT.

Identifying Hazards to Infrastructure

For the enhanced security and increased resilience of urban areas, the merits, deficiencies, and interactions of infrastructure elements need to be properly evaluated using a risk-informed approach. All hazards need to be identified, appropriate data collected, and models and methodologies developed to allow comprehensive analyses to understand risk. The complexity of our infrastructure systems of systems increases as more demands are made on infrastructure and as more infrastructure is placed underground. It is possible to model individual systems and their functionalities, but there is little certainty that models are faithful to reality. Uncertainty in behaviors of interdependent systems is relatively high, and integrated and validated models of systems of systems have not yet been developed. Our systems lose robustness as existing systems are operated more frequently and at closer to full capacity. More frequent slowdowns or shutdowns of infrastructure (e.g., traffic jams, electricity brown- or black-outs) are the result. Even though the amount of sensing and control has increased (e.g., through supervisory control and data acquisition [SCADA] systems), these systems are more vulnerable to intentional attack, unexpected failures, and loss of service (Hildick-Smith, 2005). Infrastructure sector managers can be surprised by the cascading evolution of problems across sectors. With this increasing appreciation of SCADA- and control-system vulnerabilities, anticipatory strategies need to be developed to investigate events and hazards. Problems are exacerbated because society has underinvested in rehabilitation of existing infrastructure systems, leading to deterioration, inadequate capacity, and lack of adaptation to new demands and challenges—all leading to increased vulnerability. There has been little systematic study of either the integrated risks posed by underground infrastructure systems, or of the contribution to risk posed by the increased use of the underground for the placement of critical systems.

In order for urban areas to be sustainable, infrastructure system design needs to account for successful long-term use and service as well as for the periodic and short-term critical responses to potential extreme natural or human-induced events. When extreme events occur, engineers and planners need to be educated and trained to consider them as test beds to understand complex systems behaviors, interdependencies among systems of systems, and validation of computational models of system performance.

Seismic Hazards

Sustainability of underground infrastructure and the broader community is dependent, in part, on the ability to be resilient to the risks associated with natural hazards including earthquakes and floods. Underground structures located in seismically active areas, for example, are subject to ground shaking and can experience failure if not properly designed (e.g., the Daikai Subway station, Japan; Nakamura et al., 1997). In general, underground structures perform well during seismic events due to the lower amplitudes of vibration experienced by buried facilities and the robustness of structure design and construction (Hashash et al., 2001). However, some characteristics such as depth and rock or soil properties can make underground structures more or less susceptible to damage. Hashash and others (2001) observe that deep tunnels seem less vulnerable to shaking than shallow tunnels and that facilities built in competent rock suffer less damage than those built in soils. Large earthquakes, however, can still cause significant damage on underground structures, especially near earthquake epicenters. Significant effort has been directed toward the development of design technologies to evaluate the seismic performance of underground structures (Hashash et al., 2001; Huo et al., 2005; see Box 3.1). In increasingly urban environments, however, more complete understanding of the seismic interactions between above- and below-ground infrastructure and the seismic effects on complex underground facility configurations is needed.

Flood Hazards

Flooding caused by excessive rains, hurricanes, and tsunamis is a concern in coastal and many low-lying areas. For example, storms in New York City during high tides can cause flooding of part of the subway system, necessitating the protection of surface access points against water level rise. A single storm, given the right circumstances, can result in a sustained storm surge of several feet above normal high tide levels and cause serious flooding and intrusion of saltwater into underground works, as was experienced during Hurricane Sandy on October 30, 2012 (see Box 3.9). Events such as these underscore the need for understanding risk to all hazards. Recovery from this event also can serve as a laboratory that can inform future infrastructure development and recovery planning and efforts. Underground engineers and urban planners have a unique opportunity to catalog their findings as they undertake recovery efforts, and they have the opportunity to rebuild and upgrade infrastructural systems to be more resilient and sustainable.

Flooding can occur as a result of other events or circumstances. Large magnitude subduction zone earthquakes, as in the case for parts of the western United States, may cause sea level rise that threaten infrastructure. Tsunamis also can have devastating impacts on the buildings and infrastructure of a coastal area as was demonstrated in the 2004 Indian Ocean earthquake and tsunami and the 2011 earthquake and tsunami in Japan. Interestingly, underground structures are better protected from water pressure and debris impacts of moving floodwaters if entrances are protected and sealed before an event. Areas not historically subject to flooding may not have been developed to mitigate the effects of flooding, and even some areas prone to flooding have underground infrastructure at risk (see Box 3.10 for a description of the cascading effects that flooding can have on urban infrastructure). The impacts of flooding on underground infrastructure require further research and study.

Hazards Associated with Climate Change and Sea Level Rise

The widespread environmental changes such as more frequent and intense storm events, sea level change, and flooding expected as a result of climate change will affect many urban areas (NRC, 2008b, 2010b, 2011a, 2012). Sustainability depends on the ability to respond and adapt to such changes. Although debate on climate change science is not a focus of this report, it is appropriate

BOX 3.9

Preliminary Lessons from Hurricane Sandy

Hurricane Sandy, the largest Atlantic hurricane on record, impacted the eastern United States from the Carolinas to Massachusetts on October 30, 2012 (e.g., USGS, 2012). Youssef Hashash, a member of the study committee, led a National Science Foundation-sponsored Geotechnical Extreme Events Reconnaissance (GEER) team to examine the behavior of underground and coastal infrastructure during and following Hurricane Sandy in Manhattan, Queens, Staten Island, and Rockaway Beach in New York, and in New Jersey. The GEER team's report will be released online (see http://www.geerassociation.org/), but Dr. Hashash shared preliminary observations with the committee regarding some of the impacts in Manhattan.

A 13-foot storm surge flooded lower Manhattan and destroyed or heavily damaged surface infrastructure and overtopped entrances to subways and other underground infrastructure not designed to protect against high water levels. All eight under-river subway tunnels were flooded. Although all underground infrastructure was inundated, many structures were successfully pumped and dry within days, allowing access for the GEER team. The structural integrity of these structures appeared sound, but auxiliary and life support systems (e.g., power and ventilation) were exposed for an extended period to highly corrosive and conductive seawater. To ensure safety, electrical systems could not be tested or used until inspected by qualified personnel. Many structures were "yellow tagged," indicating infrastructure owners still awaited electrical inspections; thus, at the time of the reconnaissance, the true extent of the damage could not be known.

There were more than 100 deaths in the United States as a result of Hurricane Sandy. Although many were related to drowning, none of the drowning victims was reported to have been found in public underground infrastructure (NY Times, 2012). In this way, the underground infrastructure was well-managed to avoid more casualties although little could be done to avoid infrastructural damage. Public underground infrastructure owners and operators assessed the impending hazards and risks associated with the storm and took

to bring attention to ways that development and use of underground space may mitigate potential climate change effects. Underground engineering may affect climate change drivers (e.g., land use, greenhouse gas emissions), and may increase the ability of urban communities to adapt to changing climate conditions.

Climate change refers to a statistically significant variation in either the mean state of the climate or its variability over an extended period, typically decades or longer, that can be attributed to either natural causes or human activity (IPCC, 2007). The National Research Council has reported on the consequences of climate change for the infrastructure and operation of U.S. transportation systems and identified five climate changes of particular importance including increases in number and frequency of very hot days and heat waves, increases in polar temperatures, rising sea levels, increases in intense precipitation events, and increases in storm intensity (NRC, 2008b).

General risks to infrastructure from climate change is well documented

appropriate action to clear the underground of occupants. Much of Manhattan is recovering as services are being restored. Subway service was restored to lower Manhattan as of December 3, 2012 (NY MTA, 2012a), but, as of December 10, 2012, tunnel service from Manhattan to Brooklyn had not been restored. The most heavily flooded subway tunnel (the Montague) was filled with seawater from "track to ceiling" for close to a mile (NY MTA, 2012b). It took several days to remove mud and debris from the tunnels once water could be pumped. Inspection teams found damage to signal relays, track switches, stop motors, and wiring. Debris washed into the tunnel, some with enough force to have bent metal, according to the MTA.

This is the first severe coastal flooding of a heavily urbanized part of the United States that depends extensively on underground infrastructure. The committee makes no determination about whether this event was in response to global changes in climate patterns; however, it acknowledges that more frequent or intense storms have been predicted as a result of expected climate change (e.g., IPCC, 2007; NRC, 2010). Seawater inundation experienced during this storm also serves as a reminder of expected sea level rise in many parts of the world (e.g., NRC, 2012). Observations and lessons learned from Hurricane Sandy could be collected to inform future urban sustainability decisions as emerging issues are identified and addressed.

Because many large urban areas are located along coasts and their infrastructure is already in place, more thought must be devoted to how to increase the resilience of urban areas to ensure sustainability—a difficult prospect given the age and deterioration of much of the this infrastructure. Resilience and sustainability of urban systems and infrastructure in light of all hazards and risks will necessarily be factors to consider during the decision making process. Urban planners must understand that it may be possible to reduce or mitigate the risks associated with high-intensity storms and sea level rise, but it is impossible to remove all risk. Engineers need to incorporate resilience and increased ability for disaster recovery (e.g., designing electrical components that can withstand prolonged exposure to seawater) into their technical decision making. All need to collaborate to understand what hazards infrastructure can safely accommodate.

(IPCC, 2007; CCAP, 2009; NRC, 2008b, 2010); however, addressing the consequences of climate change through better use of underground space for a more sustainable future has not been extensively studied. Further, climate change effects will vary regionally as a result of variability in natural and anthropogenic factors, so no solutions to emerging issues may be universally applicable. Issues may include the need for redesign of coastal wastewater discharge systems, better protection against flooding for underground road and rail systems, and better protection for underground utility vaults and tunnels. Making the necessary adaptations for the design of new systems and in a planned manner for existing systems will be important for the continued effective functioning of these systems. Known risks and the means to mitigate, reduce, or transfer risk will need to be considered. Options such as relocation or migration of urban centers away from areas susceptible to environmental changes could be considered but are not addressed in this report. Rethinking the placement of critical services—such as

BOX 3.10

Flooding in New Orleans Following Hurricane Katrina

The impacts of Hurricane Katrina on New Orleans provide a recent lesson regarding resilience and are discussed in detail elsewhere (e.g., Colten et al., 2008). Flooding due to levee system failure set in motion cascading failures and extensive damage to physical and social systems from which the city has not fully recovered years later. Because the pumping station used to pump storm water was not itself protected from flooding, it had to be shut down, dewatered, and dried before operations could start. Houses and buried infrastructure lines became buoyant during flooding (NIST, 2006), in many cases causing the severing of buried utility services (especially gas and water) at entry points into buildings. This created so many leaks in water and gas supply systems that supply pressures were lost and piping systems filled with unsanitary and salty water (NIST, 2006). The loss of water supply affected fire-fighting abilities and greatly slowed the return of normal living conditions. Flooding

emergency generators and fuel—in basements or flood-prone areas would be prudent. Other consequences of climate change may be unknown and warrant exploration. For example could the impacts of rising sea level on underground infrastructure include increased incidence of waterborne disease or the inability to supply water at sufficient pressure for fire-fighting during a disaster?

Some problems may emerge from placement of infrastructure underground, but the underground may offer some solutions. Questions related to underground construction, for example, include (a): can underground construction, e.g., through reduced fossil fuel consumption and carbon output, be a means to decrease human contribution to climate change? and (b) can underground construction mitigate damage or risk from environmental changes resulting from climate change? The first question involves a series of complex national or global evaluations, including calculation of the lifecycle net energy efficiency and carbon footprint of underground infrastructure versus surface counterparts (this will be discussed further in Chapter 5).

The second question regarding damage and risk mitigation relates to the use of underground space as a physical means to protect against some of the consequences of climate change, such as heavy storms, floods, and sea level rise (Bobylev, 2009). Although unprotected underground facilities can be inundated during floods, they offer increased protection against structural damage caused by water surge and debris impact. Changes in structural forces on buried facilities during storm or flood events are predictable and can be accommodated during design. It may be possible to avoid flooding by raising or protecting entrances to exclude the possibility of water ingress. Sea level rise associated with climate change poses significant risk to underground infrastructure. Global sea levels are projected to rise 8-23 cm by 2030 relative to 2000 levels, and 50-140 cm by 2100 (NRC, 2012). Some systems under construction are being designed in anticipation of future water levels. The difficulty, however, of protecting a whole

of the low-pressure gas distribution system caused corrosion of valves and meters and required extensive replacement. Shallow-buried utility lines were damaged by tree root systems when mature trees were blown over during the storm. Heavy cleanup equipment often damaged hydrants hidden by debris, and shallow-buried utilities were often driven over by such equipment, causing collapse or damage to those utilities. The lack of good or accessible records of utility line, shut-off valve, and other infrastructure element locations hampered utility and emergency services response. In addition, many normal landmarks for locating services were obliterated by hurricane damage and flooding. Recovery was slowed by the loss of urban services such as power, fresh water, and sanitation—people could not easily return to their neighborhoods even once flooding had receded. Without the residents there to clean up, many administrative and legal issues arose concerning interfaces between personal and emergency response service responsibilities (U.S. Executive Office of the President, 2006).

low-lying city from rising sea levels is daunting, examples of which can be found in the Netherlands and New Orleans, Louisiana. The low points of land in the Netherlands and New Orleans are 6.8 m and 1.5-3 m below mean sea level, respectively (Burkett et al., 2003). Underground facilities may require, among other things, special design (e.g., entrances) to make them suitable for sea level rise conditions.

Another potential engineered use of the underground in need of greater evaluation is the isolation of energy-related waste products within geologic features. The injection of carbon dioxide into geologic features for the purpose of carbon sequestration (NETL, 2010) and the isolation of high-level radioactive wastes (McCombie, 2003) are methods being studied for reliability, potential risk to people and the environment over the short and long terms, and interference with other potential underground applications. Sequestration of carbon dioxide is intended to decrease the amount of carbon dioxide—a greenhouse gas—released into the atmosphere. Underground isolation of high-level nuclear waste generated from nuclear-fission-produced electricity may indirectly reduce greenhouse gas emissions because such energy production does not result directly in greenhouse gas emissions. If the political and technical issues surrounding underground isolation of waste can be resolved, or if self-contained underground nuclear plants (each with its own long-term underground storage) were able to minimize the political, transport, and risk factors associated with both nuclear plants and waste storage (McCombie, 2003), reassessment of planning as it relates to climate change could be justified. Such issues are yet to be addressed but are outside the scope of this report.

To what heights of sea level rise is it practical to protect cities with walls and levees? Is it reasonable for threatened cities to consider abandoning existing ground floor levels, essentially raising "ground level" up one story as has been done for various reasons in parts of Seattle, Washington (Richard, 2008)?

If this occurred then the existing ground levels could become new levels of pseudo underground space, as has been accomplished in *La Defense* and *La Rive Gauche* in Paris, France (Duffaut, 2006) and Tsukuba Science City in Japan (Dearing, 1995) to create improved service infrastructure coupled with a more pedestrian-friendly environment. Given such scenarios, underground engineering technologies could assess whether existing underground pipe and cable networks could withstand additional depths of burial or flood pressure loading, how existing building basements might be reinforced against such load increases, and the potential for increased corrosion, among other characteristics.

The potential impacts of climate change on inland cities and communities could also be significant. For example, climate change–induced natural hazard events creating high-intensity rainfall activity will require system designs that capture and convey larger volumes of water to reduce or avoid flood damage and economic loss. Changes in annual rainfall will likely impact regional groundwater tables, causing changes in available groundwater supplies. Impacts on existing underground and surface structures caused by changing groundwater levels and the resulting changes in the properties of soil, rock, and materials used in underground construction also may be likely. A long-term and regional view of water management likely will be a key element in establishing resilience for local areas from such climate change effects, as will a more complete understanding of soil, rock, and construction material behavioral changes caused by changing groundwater conditions. Insurance and reinsurance, as a component of risk management of climate change events for underground systems, likely will be necessary because, although some events may have a low probability of occurrence, the consequences of their occurrence can have far-reaching spatial (geographic) and temporal economic impacts. Short- and long-term performance and infrastructure maintenance requirements will have to be understood in order to enhance resilience and sustainability.

Increasing Resilience

As discussed in Chapter 1, resilience represents the ability to respond and adapt to change in the environment. In this discussion, resilience includes the ability of an urban community to mitigate the intensity and spatial distribution of damage caused by extreme events or long-term environmental changes (for example, economic recession, climate change). The ability to respond and deliver service functionality quickly following extreme events, and to reduce economic impacts caused by the events, are demonstrations of resilience. Building resilience applies to all manner of hazards already discussed and requires removing or minimizing vulnerabilities in essential systems that place the systems at risk. It requires a system of systems approach and consideration of cross-systems interdependencies to avoid the cascading failures of individual systems.

Disasters such as Hurricane Katrina (see Box 3.10) can yield some good if society can learn from experience. How, for example, can underground infrastructure be designed to mitigate buoyancy effects as occurred during the flooding of New Orleans? How can the effects of corrosion of physical infrastructure be avoided? Chapter 2 described aspects of cascading failure caused by the collapse of the World Trade Center (WTC) towers following the terrorist attack of September 11, 2001. The attacks were tragic, but there are lessons from which planners and future responders can learn to apply to underground infrastructure design and operation:

- Con Edison Company of New York (electric, natural gas, and steam providers to New York City) used trailer-mounted portable generators to provide spot power and routed temporary feeder lines—called shunts—belowground to connect live to dead networks and restore power (O'Rourke et al., 2003; Mendonca and Wallace, 2005).
- Redundancy in subway system lines meant that access to most areas was restored in a few days (O'Rourke et al., 2003).
- Core stair systems in the World Trade Center towers resulted in evacuation routes from high in the towers that were discontinuous or became severed (underground infrastructure escape routes can suffer similarly).
- The hazards of dust on air and water quality were not immediately appreciated and were ultimately proven to be health hazards for first responders.
- A lack of readily available engineering information related to the World Trade Center towers and foundations hindered the ability to assess the potential for building collapse and stability of the foundation wall system.

The importance of the robustness of individual systems to overall resilience is highlighted by the above examples. Perhaps more importantly, the interdependencies among whole system of systems—social, economic, information, and physical systems are exposed.

Resilience of urban design depends on a multihazard approach to disaster preparation and integrated system design. A mulithazard approach necessitates planning for the most likely risk scenarios and includes enough flexibility to accommodate the unexpected (e.g., NRC 2011b). Integrated and coordinated systems planning includes the need to plan for critical redundancies in systems that, for example, allow adequate response and recovery when part of a system fails. Surface and subsurface infrastructure assets need to be designed and operated as integrated systems with lifecycle maintenance, risk, reliability, and real-time responsiveness in mind. Urban planners and engineers need trusted and validated risk-informed approaches to project planning and design that can balance project needs in terms of service delivery, initial cost, resilience against extreme events, and effective maintenance and operations so that whole life performance is satisfactory. Through adoption of this type of approach for underground space and infrastructure (occupied or not), the consequences of extreme events can be greatly curtailed and, as a result, society will widely appreciate the underground as an increasingly reliable and secure resource and part of a sustainable society.

REFERENCES

AAPA (American Association of Port Authorities). 2011. North America-Container Port Traffic: 1990-2010. Port Industry Statistics [online]. Available: http://aapa.files.cms-plus.com/PDFs/CONTAINER%20TRAFFIC%20NORTH%20AMERICA%201990%20-%202010%20for%20the%20web.pdf (accessed September 20, 2011).

ACTA (Alameda Corridor Transportation Authority). 2012a. Alameda Corridor Factsheet [online]. Available: http://www.acta.org/projects/projects_completed_alameda_factsheet.asp (accessed April 8, 2012).

ACTA. 2012b. About ACTA, History [online]. Available: http://www.acta.org/about/history.asp (accessed April 8, 2012).

APTA (American Public Transportation Association). 2008. Public Transportation Reduces Greenhouse Gases and Conserves Energy: The Benefits of Public Transportation. http://www.apta.com/resources/reportsandpublications/Documents/greenhouse_brochure.pdf (accessed June 25, 2012).

APTA. 2009. Changing the Way America Moves: Creating a More Robust Economy, a Smaller Carbon Footprint, and Energy Independence [online]. Available: http://www.apta.com/resources/reportsandpublications/Documents/america_moves_09.pdf (accessed April 8, 2011).

APWA (American Public Works Association). 1971. Feasibility of Utility Tunnels in Urban Areas. Special Report No. 39. February. Chicago, IL: APWA.

Arnold, R.L. 1992. Special Report: Underground flood hits Chicago's Loop, shutting down businesses for weeks. Disaster Recovery Journal [online]. Available: http://www.drj.com/special/chicago.html (accessed April 4, 2011).

Baron,T., G. Martinetti, and D. Pepion. 2011. Carbon Footprint of High Speed Rail Lines. Paris: International Union of Railways [online]. Available: http://uic.org/IMG/pdf/hsr_sustainability_carbon_footprint_final.pdf (accessed May 18, 2012).

Besner, J. 2002. The Sustainable Usage of the Underground Space in Metropolitan Area. ACCUS 2002 International Conference, November 14-16, 2002, Torino, Italy [online]. Available: http://www.ovi.umontreal.ca/documents/ovi_jbesner2.pdf (accessed April 7, 2011).

Bobrick, B. 1981. Labyrinths of Iron: A History of the World's Subways. New York: Newsweek Books.

Bobylev, N. 2009. Urban Underground Infrastructure and Climate Change: Opportunities and Treats. 5th Urban Research Symposium [online]. Available: http://siteresources.worldbank.org/INTURBANDEVELOPMENT/Resources/336387-1256566800920/6505269-1268260567624/Bobylev.pdf (accessed May 24, 2012).

Brueckner, J. 1999. Property Taxation and Urban Sprawl. University of Illinois at Urbana, Champaign, IL [online]. Available: http://igpa.uillinois.edu/system/files/WP80-taxsprwl.pdf (accessed September 22, 2011).

Burkett, W.R., D.B. Zilkoski, and D.A. Hart. 2003. Sea-Level Rise and Subsidence: Implications for Flooding in New Orleans, Louisiana [online]. Available: http://www.nwrc.usgs.gov/hurricane/katrina_rita/Sea-Level-Rise.pdf (accessed May 30, 2012).

Buzbee, J. 2011. Business goes underground. Progressive Engineer, May/June 2011. [online]. Available:http://www.progressiveengineer.com/features/businessUnderground.htm (accessed June 13, 2011).

CA Water Resources Control Board. 2010. Salinity. Groundwater Information Sheet, March 2010 [online]. Available: http://www.waterboards.ca.gov/water_issues/programs/gama/docs/coc_salinity.pdf (accessed June 20, 2012).

Carver, A.D., and J.E. Yahner. 1997. Defining Prime Agricultural Land and Methods of Protection. Agronomy Guide AY-283. Purdue University Cooperative Extension Service [online]. Available: http://www.extension.purdue.edu/extmedia/AY/AY-283.html (accessed June 4, 2012).

cbs2chicago.com. 2007. 1992 Loop Flood Brings Chaos, Billions in Losses, April 14, 2007 [online]. Available: http://web.archive.org/web/20070927231222/http://cbs2chicago.com/vault/local_story_104140940.html (accessed April 4, 2011).

CCAP (Center for Clean Air Policy). 2009. Ask the Climate Question: Adapting to Climate Change Impacts in Urban Regions, A. Lowe, J. Foster, and S. Winkelman, eds. Center for Clean Air Policy Urban Leaders Adaptation Initiative, Washington, DC [online]. Available: http://www.ccap.org/docs/resources/674/Urban_Climate_Adaptation-FINAL_CCAP%206-9-09.pdf (accesed May 18, 2012).

Colten, C.E., R.W. Kates, and S.B. Laska. 2008. Community Resilience: Lessons from New Orleans and Hurricane Katrina. Community & Regional Resilience Initiative (CARRI) Report 3 [online]. Available: http://biotech.law.lsu.edu/climate/docs/a2008.03.pdf (accessed July 9, 2012).

Dames and Moore's Earthquake Engineering Group. 2004. The Loma Prieta Earthquake: Impact on Lifeline Systems. Disaster Recovery World [online]. Available: http://www.drj.com/drworld/content/w1_113.htm (accessed May 14, 2012).

Dearing, J.W. 1995. Growing a Japanese Science City: Communication in Scientific Research. London: Routledge.

Duffaut, P. 2006. Underground City-Planning: A French Born Concept for Sustainable City of Tomorrow. International Symposium on Utilization of Underground Space in Urban Area, November 6-7, 2006, Sharm El-Sheikh, Egypt [online]. Available: http://uww.ita-aites.org/fileadmin/filemounts/UWW_10/use_07_underground_planning/underground_city_planning.pdf (accessed June 8, 2012).

Einstein, D. 2004. A City Reunited. Bechtel Company Magazine [online]. Available: http://www.bechtel.com/a_city_reunited_51.html [accessed May 16, 2012].

EPA (U.S. Environmental Protection Agency). 2005. National Management Measures to Control Nonpoint Source Pollution from Urban Areas. EPA 841-B-05-004. U.S. Environmental Protection Agency, November [online]. Available http://water.epa.gov/polwaste/nps/urban/index.cfm (accessed July 9, 2012).

FAO (Food and Agriculture Organization of the United Nations). 2011. Innovations in Water Management Needed to Sustain Cities. FAO Media Center, March 3, 2011 [online]. Available: http://www.fao.org/news/story/en/item/53479/icode/ (accessed June 20, 2012).

FHWA (U.S. Federal Highway Administration). 2011. SR 99: Alaskan Way Viaduct Replacement Project: Record of Decision. FHWA-WA-EIS-04-01-F. August. Available: http://www.wsdot.wa.gov/Projects/Viaduct/library-environmental.htm.

FHWA/PANYNJ (The Port Authority of NY and NJ). 2010. Cross Harbor Freight Program: Draft Scoping Document: Tier 1 Environmental Impact Statement [online]. Available: http://www.panynj.gov/about/pdf/DRAFT-Scoping-Document-Appendices.pdf (accessed May 17, 2012).

Garber, A. 2009. Alaskan Way Viaduct Legislation Is Headed for the Governor. Seattle Times, April 25, 2009 [online]. Available: http://seattletimes.nwsource.com/html/politics/2009119534_viaduct25m0.html (accessed April 12, 2011).

Goff, Z. A. 2001. Feasibility of Tube Transportation to Relieve Highway Congestion. Ph.D. dissertation. The Texas A & M University.

Greene, R.P. 2006. Strong downtowns and high amenity zones as defining features of the 21st century metropolis: The case of Chicago. Pp. 50-74 in Chicago's Geographies: Metropolis for the 21st Century, R.P. Greene, M.J. Boumann, and D. Grammenos, eds. Washington, DC: Association of American Geographers [online]. Available: http://immigrationseminar.uchicago.edu/events/GreeneChicagoGeog.pdf.

Hagler, Y., and P. Todorovich. 2009. Where High Speed Rail Works Best. America2050, September 2009 [online]. Available: http://www.america2050.org/pdf/2050_Report_Where_HSR_Works_Best.pdf (accessed September 20, 2011).

Hashash, Y.M.A., J.J. Hook, B. Schmidt, and J.I.C. Yao. 2001. Seismic design and analysis of underground structures. Tunnelling and Underground Space Technology. 16:247-293.

Hildick-Smith, A. 2005. Security for Critical Infrastructure SCADA Systems. Bethesda, MD: SANS Institute [online]. Available: http://www.sans.org/reading_room/whitepapers/warfare/security-critical-infrastructure-scada-systems_1644 (accessed July 9, 2012).

Hooks, J.M., R.D. Goughnour, W.G. Horn, A.S. Peters, E.S. Smith, G.J. Tamaro, and C.E. Thunman. 1980. A Report of the Design and Construction of Diaphragm Walls in Western Europe, 1979. Report No. DOT-FH-11-8893. Washington, DC: Federal Highway Administration, U.S. Department of Transportation [online]. Available: http://trid.trb.org/view.aspx?id=166325 (author's summary description of the report; accessed April 11, 2011).

Hricko, A.M. 2006. Ships, trucks, and trains: Effects of goods movement on environmental health. Environmental Health Perspectives. 114(4):A204-A205.

Hull, L. 2010. Now water chaos hits England as taps run dry for 3,000 after burst pipe. Daily Mail, December 31, 2010 [online]. Available: http://www.dailymail.co.uk/news/article-1343205/WATER-CRISIS-Now-chaos-hits-England-taps-runs-dry-3-000-burst-pipe.html (accessed March 9, 2011).

Hulme, T.W., and J. Zhao. 1999. Underground space development in Singapore: The past, present and future. Tunnelling and Underground Space Technology. 14(4):407.

Huo, H., A. Bobet, G. Fernandez, and J. Ramirez. 2005. Load transfer mechanisms between underground structure and surrounding ground: Evaluation of the failure of the Daikai station. Journal of Geotechnical and Geoenvironmental Engineering 131(12):1522-1533.

Hunt Midwest. 2009. SubTropolis Benefits [online]. Available: http://www.huntmidwest.com/subtropolis/benefits.html (accessed May 21, 2012).

Inouye, R.R., and J.D. Jacobazzi. 1992. The Great Chicago Flood of 1992. Civil Engineering—ASCE 62(11):52-55.

IPCC (Intergovernmental Panel on Climate Change). 2007. Climate Change 2007: Synthesis Report. Geneva, Switzerland: Intergovernmental Panel on Climate Change [online]. Available: http://www.ipcc.ch/publications_and_data/publications_ipcc_fourth_assessment_report _synthesis_report.htm (accessed June 20, 2010).

Jehanno, A., D. Palmer, and C. James. 2011. High Speed Rail and Sustainability. International Union of Railways [online]. Available: http://uic.org/IMG/pdf/hsr_sustainability_main_study_final.pdf (accessed May 11, 2012).

Lake, J.D. 1998. NFPA 520: A new standard for subterranean spaces. NFPA Journal (Nov/Dec 1998) [online]. Available: http://findarticles.com/p/articles/mi_qa3737/is_199811/ai_n8816995/ (accessed June 13, 2011).

Ledbury, M., and A. Veitch. 2012. The Benefit of Expanding Europe's High-Speed Rail Network. Public Service Europe, March 4, 2012 [online]. Available: http://www.publicserviceeurope.com/article/1589/the-benefits-of-expanding-europes-high-speed-rail-network (accessed May 17, 2012).

Lindblom, M., and E. Heffter. 2009. Alaskan Way Viaduct Tunnel Claims: Who's Right? Seattle Times, October 17, 2009 [online]. Available: http://seattletimes.nwsource.com/html/politics/2010081563_tunnelclaims17m.html. (accessed April 12, 2011).

Liu, H. 2000. Pneumatic Capsule Pipeline—Basi9999c Concept, Practical Considerations, and Current Research. Proceedings of the Mid-Continent Transportation Symposium 2000, Iowa State University, Ames, Iowa. May 5-16, 2000 [online]. Available http://www.ctre.iastate.edu/pubs/midcon/liu.pdf (accessed October 31, 2012).

Liu, H. 2004. Feasibility of Underground Pneumatic Freight Transport in New York City. Prepared for the New York State Energy Research and Development Authority. Columbia, MO: Freight Pipeline Company. 99 pp [online]. Available: http://www.uta.edu/ce/cuire/UPFT%20NY.pdf (accessed October 31, 2012).

McCombie, C. 2003. International perspectives on the reprocessing, storage, and disposal of spent nuclear fuel. The Bridge 33(3):5-10.

Meehan, R.L. 1993. A natural history of underground fuel tank leakage. Environmental Claims Journal 5(3):339-340.

Mendonça, D., and W.A. Wallace. 2005. Adaptive Capacity: Electric Power Restoration in New York City Following the 11 September 2001 Attacks [online].Available: http://www.resilience-engineering.org/REPapers/Mendonca_Wallace.pdf (accessed June 4, 2012).

Nadis, S. 2010. SubTropolis, USA. Atlantic Magazine, May 2010 [online]. Available: http://www.theatlantic.com/magazine/archive/2010/05/subtropolis-usa/8033/ (accessed May 21, 2012).

Nakamura, S., J. Esaki, I. Suetomi, N. Yoshida, and M. Iwafuji. 1997. Investigation, analysis and restoration of the collapsed Daikai Subway Station during the 1995 Hyogoken Nanbu Earthquake. Pp. 367-376 in Geotechnical Engineering in Recovery from Urban Earthquake: Proceedings of the Third Kansai International Geotechnical Forum on Comparative Geotechnical Engineering (KIG-Forum '97), January 1997, Kobe. Kansai Branch of Japanese Geotechnical Society.

NETL (National Energy Technologies Laboratory). 2010. DOE/NETL Carbon Dioxide Capture and Storage RD&D Roadmap. U.S. Department of Energy, National Energy Technologies Laboratory. December 2010 [online]. Available: http://www.netl.doe.gov/technologies/carbon_seq/refshelf/CCSRoadmap.pdf (accessed July 9, 2012).

Newman, P., and J.R. Kenworthy. 1999. Sustainability and Cities: Overcoming Automobile Dependence. Washington DC: Island Press.

NIST (National Institute of Standards and Technology). 2006. Performance of Physical Structures in Hurricane Katrina and Hurricane Rita: A Reconnaissance Report. NIST Technical Note 1476. Gaithersburg, MD: National Institute of Standards and Technology [online]. Available: http://www.bfrl.nist.gov/investigations/pubs/NIST_TN_1476.pdf (accessed July 9, 2012).

NRC (National Research Council). 2008a. Minerals, Critical Minerals, and the U.S. Economy. Washington, DC: The National Academies Press.

NRC. 2008b. Potential Impacts of Climate Change on U.S. Transportation: Special Report 290. Washington, DC: The National Academies Press.

NRC. 2010. Adapting to the Impacts of Climate Change. Washington, DC: The National Academies Press.

NRC. 2011a. America's Climate Choices. Washington, DC: The National Academies Press.

NRC. 2011b. Building Community Resilience through Private-Public Collaboration. Washington, DC: The National Academies Press.

NRC. 2012. Sea-Level Rise for the Coasts of California, Oregon, and Washington: Past, Present, and Future. Washington, DC: The National Academies Press.

NY MTA (New York Metropolitan Transit Authority). 2012a. R Resumes Service to Lower Manhattan [online]. Available: http://www.mta.info/nyct/service/RestoringRServiceMontagueStTube.htm (accessed December 12, 2012).

NY MTA. 2012b. Restoring South Ferry Station [online]. Available: http://www.mta.info/nyct/service/RestoringSouthFerryStation.htm (accessed December 12, 2012).

NY Times. 2012. Mapping Hurricane Sandy's Deadly Toll [online]. Available http://www.nytimes.com/interactive/2012/11/17/nyregion/hurricane-sandy-map.html (accessed December 12, 2012).

O'Rourke, T.D., A.J. Lembo, and L.K. Nozick. 2003. Lessons learned from the World Trade Center disaster about critical utility systems. Pp. 269-290 in Beyond September 11th: An Account of Post-Disaster Research, J.L. Monday, ed. Boulder, CO: Natural Hazards Research and Applications Information Center [online]. Available: http://www.colorado.edu/UCB/Research/IBS/hazards/publications/sp/sp39/sept11book_ch10_orourke.pdf (accessed June 4, 2012).

Palaniappan, M., S. Prakash, and D. Bailey, 2006. Paying with Our Health: The Real Cost of Freight Transport in California. A Ditching Dirty Diesel Collaborative Report, by the Pacific Institute. Hayward, CA: Alonzo Printing Co., Inc. [online]. Available: http://www.pacinst.org/reports/freight_transport/PayingWithOurHealth_Web.pdf (accessed June 20, 2012).

Parker, H.W. 2004. Underground Space: Good for Sustainable Development, and Vice Versa. International Tunnelling Association (ITA) Open Session World Tunnel Congress, May 2004, Singapore [online]. Available: http://www.ita-aites.org/fileadmin/filemounts/general/pdf/ItaAssociation/ITAEvents/OpenSessions/HParker.pdf (accessed April 30, 2012).

Parker, H.W., and J. Reilly. 2009. Life Cycle Cost Considerations Using Risk Management Techniques. Harvey Parker & Associates, Inc., Bellevue, WA, and John Reilly Associates International, Framingham, MA [online]. Available: http://www.ctta.org/FileUpload/ita/2009/papers/O-01/O-01-08.pdf (accessed April 11, 2011).

Peila, D., and S. Pelizza. 1995. Civil reuse of underground mine Openings: A summary of international experience. Tunnelling and Underground Space Technology 10(2):179-191.

PNSN (Pacific Northwest Seismic Network). 2002. Nisqually Earthquake, February 28, 2001, Basic Information [online]. Available: http://old.pnsn.org/SEIS/EQ_Special/WEBDIR_01022818543p/welcome.html (accessed May 18, 2012).

Port of Miami Tunnel. 2010. Project Overview [online]. Available: http://www.portofmiamitunnel.com/project-overview/project-overview-1/ (accessed April 8, 2011).

Richard, C. 2008. The History of the Seattle Underground. Helium.com, November 13, 2008 [online]. Available: http://www.helium.com/items/1236450-the-history-of-the-seattle-underground (accessed July 9, 2012).

Riera, P., and J. Pasqual. 1992. The importance of urban underground land value in project evaluation: A case study of Barcelona's utility tunnel. Tunnelling and Underground Space Technology 7(3): 243-250.

Roop, S.S., C.E. Roco, L.E. Olson, C.A. Morgan, J.E. Warner, D.-H. Kang. 2003. Year 4 Report on the Technical and Economic Feasibility of a Freight Pipeline System in Texas. Report No. FHWA/TX-04/9-1519-4. Texas Transportation Institute, Texas A & M University System.

Schofield, J. 2011. London after the Great Fire. BBC History [online]. Available: http://www.bbc.co.uk/history/british/civil_war_revolution/after_fire_01.shtml (accessed May 21, 2012).

Sterling, R., H. Admiraal, N. Bobylev, H. Parker, J.P. Godard, I. Vähäaho, C.D.F. Rogers, X. Shi, and T. Hanamura. 2012. Sustainability issues for underground spaces in urban areas. Proceedings of ICE - Urban Design and Planning [online]. Available: http://www.icevirtuallibrary.com/content/article/10.1680/udap.10.00020.

Uffink, T., and H. Admiraal. 2012. Underground Transport of Cargo: A Sustainable Alternative. Proceedings of the 13th World Conference of Associated Research Centers for the Urban Underground Space, Marina Bay Sands, Singapore. November 7-9, 2012. CD ROM (accessed October 31, 2012).

UrbanRail.Net. 2007. Subway Maps: Duisburg [online]. Available: http://www.amadeus.net/home/subwaymaps/en/info/duisburg.htm (accessed April 22, 2012).

U.S. Executive Office of the President. 2006. The Federal Response to Hurricane Katrina: Lessons Learned. Washington, DC: Government Printing Office.

USGS (United States Geological Survey). 1998. The Loma Prieta, California, Earthquake of October 17, 1989—Recovery, Mitigation, and Reconstruction, J.M. Nigg, ed. U.S Geological Survey Professional Paper No. 1553-D [online]. Available: http://pubs.usgs.gov/pp/pp1553/pp1553d/pp1553d.pdf (accessed May 14, 2012).

USGS. 2009. Historic Earthquakes: Santa Cruz Mountains (Loma Prieta), California [online]. Available: http://earthquake.usgs.gov/earthquakes/states/events/1989_10_18.php (accessed November 23, 2010).

USGS. 2012. Coastal Change Hazards: Hurricanes and Extreme Storms [online]. Available: http://coastal.er.usgs.gov/hurricanes/sandy/ (accessed December 12, 2012).

Weston Solutions. 2005. Alameda Corridor Air Quality Benefits Final Report: June 10, 2005 [online]. Available: http://www.acta.org/pdf/Alameda%20Corridor_AQ_Benefits_061005.pdf (accessed July 18, 2012).

Wren, J. 2007. The Great Chicago Flood: Analysis of Chicago's 2nd great disaster. Structure Magazine (August):35-40 [online]. Available: http://www.structuremag.org/Archives/2007-8/SF-Chicago-Flood-Wren-Aug-07.pdf (accessed April 4, 2011).

WSDOT (Washington State Department of Transportation). 2004. SR 99: Alaska Way Viaduct & Seawall Replacement Project: Draft Environmental Impact Statement. March 2004 [online]. Available: http://www.wsdot.wa.gov/Projects/Viaduct/library-environmental.htm#deis [accessed May 14, 2012].

4

Health and Safety Underground

> "*It is important to note the significance of human behavior on tunnel safety. The final outcome of some incidents may depend more on the quick and right reaction of individuals than on the technical safety level in the tunnel.*" (OECD, 2006, p. 15)

We design, build, and operate all manner of underground infrastructure, but doing so must occur with due consideration of the abilities and behaviors of the underground infrastructure operators and occupants to minimize risks and increase efficiencies (see Box 2.1). George Bugliarello described the need to balance the human and mechanical elements of urban living to create modern, environmentally sustainable, and emotionally satisfying environments (Bugliarello, 2001). Safety is also a necessary part of this vision. Underground infrastructure systems are complex and have elements similar to what Bugliarello described as biosoma systems—systems that include biological (individuals that create, manage, or use the system), social (organizational aspects), and machine components (the engineered artifacts). Bugliarello acknowledged the interfaces of these elements in transportation systems to be points of vulnerability that ultimately impact system resilience (Bugliarello, 2009). This committee contends the same could be said more generally as humans move into the underground where the infrastructure will be critical to support this movement.

Taking the idea further, it can be said that urban sustainability is as much dependent upon human activities, ideas, and behaviors as it is upon the robustness and resilience of physical infrastructure. Resilience of a community is tied to the resilience of physical infrastructure (e.g., Miles, 2011), but an understanding by the people who design, operate, use, or benefit from underground infrastructure of the role each structure and system element serves in the proper functioning of the urban system is important to address the robustness, resilience, and sustainability of the urban system.

Real hazards and risk to humans in the underground exist, and engineers have been largely successful in addressing many of them. Earlier chapters of this report looked at how urban utilities and systems are highly integrated and therefore interdependent. This chapter addresses human-technical system relationships, human response to hazards faced in the underground, and the hazards and risks related to human use of underground space. This chapter recognizes the people in the underground and considers the engineering necessary to keep them healthy while also contributing to sustainability. The presence or absence of naturally occurring phenomena in the underground may pose risk to humans. Gases, radiation, temperature, water, and the lack of oxygen are among inherent hazards to human underground occupation. Other hazards to people or infrastructure may result from human activity that creates, adds to, or intensifies naturally occurring risks. These include risks associated with fire and smoke, hazardous materials, intentional or accidental explosions, structural failure, human failure, and extreme events.

It is important to fully understand the hazards and risks because a very key part of long-term success (i.e., sustainability) of the underground is the ability to regulate underground construction and activities to ensure minimum safety. Although various standards exist that govern, principally, fire safety for underground transportation and building and industrial facilities, there is a need for a more comprehensive approach to safety against all hazards for all types of underground facilities. The remainder of this chapter explores this need.

HUMAN FACTOR ENGINEERING

To create a functioning, sustainable, urban system that effectively links its social, technical, and governing elements, the relationships between technologies, the people that construct, operate, and use those technologies, and the social structures that govern them must be understood. In the manufacturing realm, this area of research is referred to by several names including human factors, human engineering, engineering psychology, and ergonomics. Licht and others (1989) analyzed numerous definitions for terms and areas of study related to or synonymous with "human factors" research and found that most definitions implied a multidisciplinary approach including concepts related to behavioral science; human performance capacity; manpower, personnel, and training; and biology, physiology, and medicine.[1] Information obtained through the study of human factors can be applied to the "design of tools, machines, systems, tasks, jobs,

[1]Biology, physiology, and medicine were more common in definitions associated with ergonomics (Licht et al., 1989).

and environments for safe, comfortable, and effective human use" (Chapanis, 1991, p. 1) so that we may "optimize the relationship between technology and the human" (Kantowitz and Sorkin, 1983; Licht et al., 1989, p. 27). The application of Complex Adaptive Systems of Systems engineering as discussed in Chapter 2 would necessarily consider the relationships between humans and underground infrastructure.

The military has long recognized the importance of integrating human and technological system elements to make operations as effective, efficient, safe, and sustainable as possible, and has promulgated these concepts through directives and guidance. For example, a Department of Defense (DOD) directive from 1988 required consideration of manpower, personnel, training, and safety in the defense system acquisition process for the purpose of improving "all aspects of the human-machine interface" (DOD, 1988: p. 1).[2] In 2007, the National Research Council published a report at the request of the Army Research Laboratory, the Air Force Research Laboratory, and DOD to address approaches for creating "an integrated, multidisciplinary, generalizable, human-system design methodology" (NRC, 2007, p. 2). That report outlines principles considered critical to human-system development and evolution including those associated with the need for stakeholder consensus on desired outcomes, regular reassessment of plans based on lessons learned, and risk management.

Many applications of human factors engineering are related to human interaction with a single manufactured item or technology. Underground systems as part of total urban environments are more complex, and the need to understand, design, regulate, and operate for human-technology relationships becomes amplified. The impact of failure of key infrastructural components—including human—and or systems can be devastating to sustainable functioning of the urban environment (see discussion of cascading failures in Chapter 2). Human behavior is not always predictable in the face of adverse and extreme events, and regardless of how resilient to hazards underground infrastructure and safety systems may be, infrastructure and system failure could have significant negative consequences. All forms of underground engineering not only must consider what training and safety guidelines are necessary for the smooth functioning of infrastructure in the best of circumstances, but also must anticipate the behavior of underground occupants during both normal and worst-case operation scenarios. Design must be holistic and create an integrated environment that allows people to almost intuitively understand how to remain safe should adverse conditions arise. Sustainability of the urban setting is dependent on optimization of human-technical relationships in ways that provide at least minimum safety while remaining consistent with long-term societal visions.

Industry also addresses safety in underground infrastructure. The International Tunneling Association (ITA), for example, established the Committee on

[2]This directive has since been replaced by other directives that also emphasize human factors.

Operations Safety of Underground Facilities (COSUF)[3] to address operational concerns of safety and security in underground structures. COSUF has developed risk assessment guidelines (Molag and Trijssenaar-Buhre, 2006) and, with an ITA working group on health and safety,[4] focused on increasing safety practices during construction. The European Construction Technology Platform (ECTP) acknowledges that safety and security must be designed into every element of infrastructure, including the interfaces between every element, with consideration of the entire life cycle of the infrastructure (ECTP, 2005).

MANAGING SAFETY THROUGH REGULATION

It may be expected that safety in underground infrastructure will be equal to that of surface infrastructure, and if not, then the expectation may be that one is fully informed of potential risks. However, although engineers have been successful in reducing many types of risk associated with underground space use, risk in underground infrastructure has not received the same level of regulatory scrutiny as risk associated with surface infrastructure, and the levels of certain risks may not be well understood. Existing codes tend to be prescriptive in nature—prescribing specific procedures or materials—but underground space poses different safety challenges that codes intended for surface space were not designed to address. For example, most people know that simply leaving a building that is on fire is adequate to reach safety. Exiting a tall building during an emergency, for example, usually requires its occupants to climb down several flights of stairs rather than use elevators or escalators. However, leaving an underground structure on fire may only move occupants to a different underground space also contaminated by smoke, and occupants may have to exit up several flights of stairs—a physically challenging task for some. Hazards associated with elevators and escalators are partially addressed by the American Society of Mechanical Engineering Safety Code for Elevators and Escalators (ASME, 2010a) that covers design, construction, installation, operation, maintenance, alteration, inspection, and testing of elevators and escalators. Guidelines also provide information on how Department of Justice requirements related to the Americans with Disabilities Act will be met by the performance of elevators or escalators (ASME, 2010b).

Safety sometimes needs to be created operationally rather than through technical solutions (e.g., no hazardous materials unless appropriate sprinkler or other systems are in place). Safety codes are most often written in response to lessons learned from incidents or litigation rather than in response to research. A responsible risk management strategy includes identifying and understanding

[3] See http://cosuf.ita-aites.org/ (accessed June 15, 2011).

[4] For example, the Health and Safety in Works working group of the International Tunneling Association has released multiple publications related to safe working practices (see ITA-AITES, 2011).

hazards and risks and applying appropriate mitigation strategies. Once recognized, underground risks may be avoided, transferred, or reduced to tolerable levels. In some cases the cost for mitigation may be substantial or prohibitive either in terms of capital costs for construction or in operational costs. This could mean a project is never started, or that minimum systems put in place may not be optimally maintained due to the costs. Assuming that avoiding or transferring risk is not feasible, reducing risk through appropriate safety regulations and education may be the best approach. Safety standards for surface infrastructure have been developed at the federal, state, and local levels and refined over generations to cover a broad array of activities. Such standards serve a key role in preventing or mitigating risks.

Current federal-level safety regulations for underground infrastructure are limited, do not apply to everyday usage of most types of facilities, and mostly are intended to regulate construction safety through the Occupational Safety Hazard Administration (OSHA). They include the OSHA regulations related to underground construction (29 CFR 1926.800)[5] that apply to construction of underground tunnels, shafts, chambers, passageways, and cut-and-cover excavations connected to underground construction to reduce hazards associated with "reduced natural ventilation and light, difficult and limited access and egress, exposure to air contaminants, fire, flooding, and explosion" (OSHA, 2003). The regulations define a tunnel as a subsurface excavation, "the longer axis of which makes an angle not greater than 20 degrees to horizontal." Although applicable to many types of underground infrastructure, the regulations are only intended to protect underground construction workers during construction and do not address safety issues once the infrastructure is in operation.

Each state in the United States has adopted fire and life safety codes to ensure safety in structures, but the codes do not fully address underground structures. Most states (45) have adopted the International Code Council (ICC)[6] building, fire, plumbing, and mechanical codes. The ICC codes refer to three National Fire Protection Association (NFPA) standards—NFPA 130 (NFPA, 2010a), NFPA 520 (NFPA, 2010b), and NFPA 502 (NFPA, 2011)—that address underground fire and life safety and that provide safety guidelines for road and passenger rail tunnels and use of space created by underground excavation. Two of these standards have been applicable to underground transportation facilities for decades. However, the applicable NFPA standards cannot adequately address underground fire and life safety for all underground space uses, and they will likely break down when combining different types of occupancy in one underground space. Additionally, the standards have limited legal authority unless adopted by states or local jurisdictions.

---

[5] See http://www.osha.gov/pls/oshaweb/owadisp.show_document?p_id=10790&p_table=STANDARDS (accessed April 4, 2011).

[6] See http://www.iccsafe.org/Pages/default.aspx (accessed June 9, 2011).

The inadequacy of safety standards results from their being developed without due consideration of the growth of all types and large scales of underground use. Innovation in underground design and construction may be bound by prescriptive (and potentially ineffective) codes when performance-based mechanisms that ensure designs will perform as intended are really needed.[7] Further, as is stressed throughout this report, underground infrastructure is only one element of the total urban system that is increasingly interconnected and interdependent. Decisions regarding safety of one infrastructural element need to be based on the effects of that decision on the overall system. Demand for underground space use is growing, and without carefully considered, research-based, national-level guidelines or effective safety standards that account for the underground as part of the larger integrated urban system, local jurisdictions are left to establish their own safety standards that may not be fully informed if appropriate resources and capacities are not available.

HAZARDS TO HUMAN HEALTH

The next sections discuss hazards to human health associated with occupying the underground, focusing on lack of adequate ventilation, smoke from fire, and hazardous materials. Some hazards and risks can be prevented operationally, others can be addressed through engineering solutions directly into infrastructure design, and others can be controlled by systems. Careful analysis of underground emergency scenarios for all hazards and risks—including those emerging as a result of changes in technologies or use—could ensure that underground emergency incidents do not escalate beyond the possibility of control or cause preventable damage. For example, the current trend toward more electric vehicles could be seen to reduce the risk of fires in tunnels, but the batteries in electric vehicles present their own set of risks. Future fleets of vehicles powered by hydrogen or natural gas present still different concerns.

Redundancy in fire and life safety systems is a key to controlling incidents. For example, because underground smoke management is critical, it is essential to ensure that the minimum ventilation scenarios to control smoke from a fire are operational if any portion of emergency ventilation fails. Without this level of redundancy in essential life safety systems, a simple mechanical failure could jeopardize the underground occupants, contents, and physical structures.

Ventilation, Smoke, and Fire Control

Underground ventilation engineering entails providing breathable air to people underground and removing hazardous gases (e.g., excess carbon dioxide,

---

[7] Asia has more performance-based codes because contracting there is design-build rather than design-bid-build, which is more common in the United States.

exhaust, fumes) from occupied space. Simply moving air from the surface to the underground may not be adequate, because the air must contain enough oxygen for the volume of people to be supplied and be free of contaminants. Hazardous gases can be removed by cleaning the underground air or by safely (to those underground and at the surface) routing contaminated air to the surface. NFPA and ICC standards address many of these issues, but not explicitly for underground spaces created for human use. Risks may be inadequately quantified.

One of the greatest hazards to human health and safety in the underground is smoke from fire (ITA, 1998). It is well within current technical knowledge and life safety system capacity to manage smoke in nearly all types of surface structures. However, managing smoke in a complex underground structure—one that can span multiple underground levels over several city blocks, be occupied by thousands of people at any given time, and has many uses (e.g., retail, office space, health care, residential) and therefore many potential hazards and risks—may challenge the most sophisticated ventilation system designs. These underground multi-use areas can be far more complex and difficult to ventilate than, for example, some roadway tunnels that can be modeled as simple tubes of air with, although long, relatively small cross-sections.

There are important strategic distinctions in the management of smoke in high-rise buildings versus large underground structures. For example, a 40-story high rise that occupies a full city block (creating the equivalent of 40 city blocks of floor space in a vertical alignment) is typically designed to control ventilation, fire sprinklers, alarms, and exiting systems immediately for up to four floors of the building where occupants are most at risk. Smoke management is typically limited to stairwell and elevator shaft pressurization that require relatively small fans. Occupants on other floors are protected by the structure's intervening floors for a short time until they can safely evacuate.

An underground structure of comparable size (the equivalent of 40 city blocks of floor space) potentially can occupy a broader lateral space over fewer levels, increasing the lateral exposure to fire and smoke that spreads throughout the horizontal space. Smoke management in such a large-scale area—with few, if any of the control tools available in buildings (e.g., windows to the outside)—requires comparatively more complex design and more powerful ventilation systems, larger sprinkler systems, and carefully designed fire detection, alarm, and exiting systems to protect occupants. Specialized emergency alarm information need to be designed to notify people of the need to evacuate. High rise building systems need only address one floor at a time. Underground systems necessarily accommodate the equivalent of 20 or more floors simultaneously.

Preventing fire and inhibiting fire growth are possible through management strategies including non-combustible construction, automatic fire suppression, precise fire detection, compartmentalizing, control of hazardous materials, heightened security, and careful occupancy restrictions (e.g., to prevent proximate hazards such as factory work adjacent to hospitals). Underground structures may have some advantages in terms of fire and life safety as compared to surface structures. Underground structures with smaller enclosed spaces may permit utilization of a fine water mist or gaseous systems to control fire and smoke, thereby reducing the demand for water and drainage which can create other problems in the underground. Simply ensuring that the occupants recognize the hazard of fire in the underground may ensure that all occupants take fewer risks associated with fire.

Hazardous Materials

Hazardous materials used in or created by manufacturing, processing, and shipping pose special risks in the underground for reasons similar to those for smoke and fire: the physical separations and ventilation systems that provide adequate safety aboveground may not be adequate below grade. On the surface, for example, a machine shop that employs cutting torches may be permitted to operate in a building next to a residential structure provided that a firebreak such as an open air gap exists between the walls of the two structures. A sufficient air gap ensures that a fire in the machine shop does not readily spread to adjoining structures, and allows easy air exchange to the outside so that gases used in cutting processes do not displace oxygen and create an oxygen-deficient atmosphere. On the other hand, engineering and operational measures may be needed to ensure safety in underground structures. Proper firewalls, ventilation, and procedures may be necessary for safe cutting in the underground. Similarly, underground spillage of hazardous liquids may pose long-lasting health risks if they migrate via underground ventilation and drainage systems or penetrate adjoining soils and porous rocks to contaminate other spaces or water supplies.

SECURITY FROM VIOLENCE

Underground infrastructure is often designed to make underground facilities attractive and easier for the public to access and use. Even underground public utilities, although not designed for access by the general public, need to be designed to accommodate access by workers and equipment. Infrastructure design often includes security elements to prevent crime and vandalism or to protect against fire or similar emergencies. Unfortunately, design elements that allow easy access to the underground by ordinary public citizens also allow access to those with dangerous or destructive intent. It is impossible to foil all attempts of violence against people or infrastructure (Jenkins and Gersten, 2001). Even so, ridership trends of underground metros in large U.S. cities have risen in the past 10 years (e.g., WMATA and Cambridge Systematics, Inc., 2009; DiNapoli and Bleiwas, 2010), indicating that need and convenience outweigh immediate concerns over personal safety for at least some percentage of the population. Few studies have documented underground use patterns following terrorist events, but studies of public transit ridership in the aftermath of the 2005 London bombings, the 2004 Madrid bombings, and the 1995 Sarin gas attacks in Japan revealed that behavior is influenced by cultural beliefs, characteristics of the attack, factors associated with the transportation system itself, and social perceptions of risk (von Winterfeldt and Prager, 2010). For example, London underground and bus (also targeted in the attack) ridership dropped but slowly recovered after the incident there, but ridership in Japan did not seem to change (Prager et al., 2010b).

Security and resilience to violence in an urban community can be enhanced through a variety of planning, design, and operational functions that reduce the frequency or severity of hazardous events. This section first discusses the safety of individuals from personal violence and then discusses violence against larger numbers of people and infrastructure itself.

Safety from Crime

A sense of personal safety—the freedom to function in a city with a low expectation of violent attacks against one's person—is important for the smooth functioning of society. The physical design of and the number of people present in an occupied space contribute to safety of individuals and the sense of personal safety. Certain types of underground structures, for example pedestrian underpasses, may have a poor reputation with respect to safety, perhaps due to poor lighting or limited occupation, as compared to metro systems where higher levels of security are in place to manage passenger organization (for example, through the use of shorter trains and platform use at night to increase the number of passengers in occupied areas). Mixed underground use offers different sorts of problems. How is the security, for example, of a retail operation located in a public transportation concourse assured when the retail space is closed for business at night but when public transportation is still in use? How is public access to transportation assured if an underground shopping area is closed for the day? Engineering solutions may come in the form of enhanced monitoring (see Chapter 6).

Attacks against Infrastructure and Urban Populations

The underground has long been and still is suggested or used for either containment or security. For example, the underground is used to protect the security of a nation's leadership (McCamley, 1998; Barrie, 2000). With the advent of weapons of mass destruction, a great deal of engineering work was done in the 1950s and 1960s on underground military and defense facilities in the United States that served to advance technologies related to the environment, security, and fire protection in underground facilities. Examples include the Cheyenne Mountain alternate command facility deep in a granite mountain and the bunker at Greenbrier in West Virginia for the continuity of government in the event of an attack. Additionally, there is continued interest worldwide in placing nuclear power plants and their waste underground to increase isolation of radioactive materials as well as to increase security of the facilities (e.g., Myers and Elkins, 2009). The feasibility of long-term storage and safety continues to be an active field of investigation. In recognition of the security offered by the subsurface, the Svalbard Global Seed Vault in Norway was constructed in a mountain to protect global crop diversity in the event of climate- or war-related regional or global catastrophe (Fowler, 2008).

The September 11, 2001, (9/11) terrorist attacks on the United States, however, fundamentally changed the way safety and security are addressed in this country, including the design and operation of underground structures. Prior to 9/11, vandalism and criminal activity were the main concerns for underground security. Terrorist threats against people and infrastructure were considered anomalies. Underground infrastructure, especially mass transit systems, is now recognized as a vulnerable target by those individual wanting to do large amounts of damage to infrastructure or to inflict harm on large numbers of people. The effects of explosions, fire, gases, and other airborne toxins and health hazards can be more concentrated and deleterious in confined underground structures. Acts of terrorism have occurred in several underground locations with serious consequences, for example, the 1995 attack with the nerve gas, Sarin, in Tokyo, Japan (Tu, 1999), the 2005 bombings in London, England (HC, 2006), and the 2010 bombings in Moscow, Russia (Rogoza and Zochowski, 2010). All of these events were perpetrated using devices carried by hand into underground infrastructure.

Approximately 87 percent of terrorist attacks around the world in 2003 were perpetrated through bombings (U.S. Department of State, 2004), which may be delivered as vehicle-borne improvised explosive devices, devices employed as booby traps, remotely detonated devices, or devices delivered by human bombers. There also is a conceivable threat of targeted ground-penetrating explosive devices delivered by missiles. However, underground installations have been recognized as providing the "most effective physical protection available" and can be designed so that critical infrastructure elements are protected against physical attack and hardened against electronic attack (Linger et al., 2002). Underground placement of facilities makes them harder to damage from the outside (i.e., from the surface) and limits points of entry. Linger and others (2002) describe the cost of that protection as "competitive" with aboveground structures hardened to similar levels. Unfortunately, classification of military technologies has resulted in a lack of standards or practices in civilian infrastructure (Gui and Chien, 2006).

Recognizing the need to address such hazards, multiple organizations have initiated research related to many aspects of underground security and safety. The American Association of State Highway and Transportation Officials (AASHTO) Transportation Security Task Force sponsored the preparation of a guide to assist transportation professionals as they identify critical highway assets and take action to reduce their vulnerability (SAIC, 2002). The Transportation Research

Board of the National Research Council has released many reports related to transportation safety and security, including many related to underground transportation.[8] These reports provide guidelines and recommendations on topics such as permanent enhancements to underground infrastructures that will improve security as well as the usable life of the underground structure and support systems (TRB, 2006). Similarly, the Federal Highway Administration and AASHTO jointly sponsored a panel to develop "strategies and practices for deterring, disrupting, and mitigating potential attacks," recommending that interagency and stakeholder coordination occur so that security assessment methodologies and solutions are consistent with needs of all involved and that federal- and state-level legal responsibilities are clarified (BRPBTS, 2003). From a technical point of view, the panel recommended that critical bridges and tunnels be identified and prioritized, and funds allocated to cover security of those structures. The panel further recommended that security should be an engineered element of design and that appropriate research and development should inform technical standards for structures in consideration of security threats.

Security, like safety, is enhanced by collaborative systems thinking among all stakeholders throughout the life cycle of the infrastructure. Interaction between urban planners and underground engineers during development and operation can focus on how underground infrastructure can improve or impede protection of critical facilities and their occupants. Security issues and needs constantly change as technologies change, known hazards are successfully addressed, or new hazards evolve. Sustainability requires applying innovative and comprehensive technologies, and, as often described in the security arena, technologies must include the concepts of prevention, deterrence, detection, and delay (e.g., Rowshan et al., 2005), as well as the concepts associated with response, recovery, and evaluation of lessons learned from incidents or "near misses" that do occur.

Massive loss of life and grave structural, economic, and even political damage may result if security threats are not appropriately assessed and addressed. Ensuring the safety of people and physical assets and minimizing disruption of the physical, social, and economic infrastructures of the total urban system must be considered. However, each underground system element is unique and requires specific measures to mitigate a range of anticipated threats. Passive hardening is, in reality, the last line of defense in providing a safe and secure facility, and passive structural hardening techniques applied to reduce vulnerability will not necessarily increase sustainability.

Introduction of human factor engineering to prevent panic and errant behavior and to guide threat recognition, decision making, and action under stress are called for. New materials and their behaviors for this application must be consid-

[8] See http://onlinepubs.trb.org/Onlinepubs/dva/CRP-SecurityResearch.pdf (accessed June 15, 2011) for a status report of cooperative research programs related to security, emergency management, and infrastructure protection.

ered (e.g., to prevent injury from fragments and flying debris and the development of airborne toxins from chemical changes due to heat and fire). In addition, the risk assessments need to include aspects of evacuation, rescue, and recovery to minimize impacts and assist in post-incident activities.

INTERNATIONAL UNDERGROUND TUNNEL SAFETY CODES

International safety codes and guidelines applicable to underground infrastructure are not enforceable in the United States, but comparison to U.S. codes can be helpful to reveal inadequacies in practice and guide future practice in the United States. The U.S. Federal Highway Administration (FHWA) sought to learn what underground systems, equipment, and procedures were employed internationally to improve underground safety, operations, and response (Ernst et al., 2006) and ultimately made recommendations for implementation strategies in nine areas in which U.S. standards and regulations could be improved (see Box 4.1).

The most comprehensive international safety information related to underground infrastructure deals with road tunnel construction and operation, and the United Nations Economic Commission for Europe (UNECE) has found that there are fewer traffic accidents in long tunnels than on similar length stretches on the open road, which is attributable to protection from the elements and consistent lighting (UNECE, 2001). However, incidents that do occur in tunnels are likely to have greater impact in terms of harm to people and infrastructure. UNECE states that improving motorist behaviors, their vehicles, tunnel operator efficiency, and the infrastructure itself are ways to decrease the number of tunnel incidents. UNECE findings are acknowledged in a directive from the European Union on minimum safety standards for tunnels in the trans-European road network (European Parliament and Council, 2004).[9] The World Road Association (PIARC)[10] is another international forum that considers an array of road and transport issues from the point of view of sustainability. Its standing technical committee is tasked with exploring management and improvement of tunnel safety, influencing user behavior in tunnels, and evaluating, organizing, and communicating knowledge on tunnel operations and safety. PIARC has produced several safety documents including those related to controlling fire and smoke in road tunnels, human factors and road tunnel user safety, and integrated approaches to road tunnel safety.

[9]All tunnels longer than 500 meters belonging to the road network are to meet minimum safety requirements related to organization, roles, and responsibilities of various administrative bodies in charge of tunnel safety, and related to technical standards for tunnel infrastructure, operation, traffic rules, and user information. Approximately 500 tunnels in Europe in operation, under construction, or at the design stage are affected. Retroactive requirements for safety are also detailed in the directive.

[10]PIARC Technical Committee. 3.3 Road Tunnel Operations. Available: http://www.piarc.org/en/Technical-Committees-World-Road-Association/ (accessed June 27, 2012).

EMERGENCY RESPONSE CHALLENGES

Response to underground emergencies of all types poses distinct challenges to emergency responders who typically develop strategic and tactical plans and train for response scenarios. Response time to underground emergencies is increased, access and way finding may be difficult, and the complex environment makes intuitive decision making more challenging. Emergency responders require specific training and practice to use the more complex fire and life safety systems that manage, for example, smoke, fire suppression, access, exiting, and fire notification in the underground.

Response Times

Fire and medical services are mandated to respond to calls as quickly and safely as possible. For example, NFPA 1710 establishes a 4-minute minimum response time by firefighters to the "front door" of the structure for 90 percent of all incidents (NFPA, 2010c). However, the "front door" of an underground structure could be its street level access portal, possibly several blocks distant from the emergency site. The distance increases the time firefighters can respond to the actual emergency. If lengthy response times are unacceptable, responders and equipment may need to be located underground or closer access points included in design. For larger underground complexes, underground emergency resources may include emergency apparatus (fire engines, ladder trucks, and medical vehicles) and law enforcement.

Accessing an underground fire or other hazard may require a difficult descent through rising smoke unless alternate access routes or methods are designed, built, and maintained. Fire fighting activities are difficult because normal processes for visual assessment of a situation on the surface, typically accomplished through inspection of at least three sides of the incident building, may not be practical underground. Emergency ventilation by vertical or horizontal methods may be limited by lack of exterior windows or access to the exterior by a 'roof' where smoke can be released to the outside.

Wayfinding

The ability of emergency responders to orient themselves is critical. Extra steps are necessary to ensure use of a comprehensive methodology that provides exact and rapidly recognizable locations. Many emergency response departments now use satellite technology to locate response units and employ computer aided dispatch (CAD) to identify the units with the shortest possible response time. However, these technologies depend on line-of-sight communication with satellites and are not functional underground. Alternatives have yet to be developed for underground use, and emergency responders must rely on old technologies

BOX 4.1

Recommendations and Implementation Strategies for Improved U.S. Tunnel Safety from the International Technology Scanning Program

The U.S. Federal Highway Administration and the American Association of State Highway and Transportation Officials and the National Cooperative Highway Research Program sponsored a study to explore practices in several European countries related to tunnel safety, operations, and emergency response. The following are recommendations and some implementation strategies excerpted from the resulting report (Ernst et al., 2006).

1. Develop Universal, Consistent, and More Effective Visual, Audible, and Tactile Signs for Escape Routes.
Recommendations include uniformity of signage that could be understood by all people and minimizes confusion in locating an exit in case of an emergency. Sounds and simple verbal messages and tactile messages could make visual signs more effective in low light situations. National Fire Protection Association (NFPA) guidelines applicable to fire protection and life safety designs should be reviewed, and current technologies and results from human response studies should be incorporated into the design of escape portals, escape routes, and cross passages (See Figure 1).

2. Develop AASHTO (American Association of State Highway and Transportation Officials) Guidelines for Existing and New Tunnels.
A single AASHTO reference for engineers and operators to facilitate consistent U.S. criteria coordinating AASHTO, FHWA, NFPA, American Public Transportation Association, and National Research Council Transportation Research Board standards and guidelines for tunnels.

3. Conduct Research and Develop Guidelines on Tunnel Emergency Management That Includes Human Factors.
Learn from European human factor design experience for more effective tunnel planning, design, and emergency response. Work through AASHTO to

FIGURE 1 Example from Mont Blanc Tunnel (between France and Italy) of tunnel escape route signage, typical of the uniform signage used throughout many countries in Europe.
SOURCE: Ernst et al., 2006.

fund and develop tunnel emergency management guidance. Collaborate with academe to study human response in tunnel incidents.

4. Develop Education for Motorist Response to Tunnel Incidents.

5. Evaluate Effectiveness of Automatic Incident Detection Systems and Intelligent Video for Tunnels.

Computer systems connected with video surveillance systems can be used to detect, track, and record incidents and signal operators to take appropriate action, decreasing response time. Because widespread public use of closed-circuit television is not readily accepted in the United States, outreach explaining the benefits and possibilities of this technology would be necessary.

6. Develop Tunnel Facility Design Criteria to Promote Optimal Driver Performance and Response to Incidents.

Innovative tunnel design with improved geometry or that is more aesthetically pleasant enhances driver safety, performance, and traffic operation.

7. Investigate One-Button Systems to Initiate Emergency Response and Automated Sensor Systems to Determine Response.

Some human errors and the need for decision making in emergency situations may be avoided with a single button for operators to press that initiates multiple critical response actions. Automated systems (e.g., using opacity sensors) may help determine appropriate responses to certain situations. Fans and vents may best be controlled through closed-loop data collection and analysis systems that monitor atmospheric conditions, tunnel air speed, and smoke density.

8. Use Risk-Management Approach to Tunnel Safety Inspection and Maintenance.

Intelligent monitoring and analyses of data can provide information to allow risk-based decision making with respect to scheduling inspections (and inspection frequency) and priorities.

9. Implement Light-Emitting Diode Lighting for Safe Vehicle Distance and Edge Delineation in Tunnels.

Blue LED lights at specific intervals allows drivers to more easily gauge distance from tunnel walls and vehicles in front and to maintain safe driving distances (see Figure 2).

FIGURE 2 Some European tunnels use evenly distributed light-emitting diodes to help motorists discern the roadway edge and determine safe following distances.
SOURCE: Ernst et al., 2006.

(e.g., hard copy maps) to orient themselves in underground settings. Whereas maps are a viable alternative, using more than one technology (e.g., satellite and hard copy maps) may create confusion for responding units. Some technologies for emergency communications and tracking that may have application in underground infrastructure are being researched and tested through support provided by National Institute for Occupational Safety and Health Office of Mine Safety and Health Research. For example, the agency supports research for inertial navigation for self tracking and wireless communication for use by miners and rescue personnel.[11] Advancement of these technologies may lead to eventual improvements in underground infrastructure safety.

Response to terrorist events in underground infrastructure can be particularly challenging for emergency responders because responders, along with key safety systems (e.g., exits), also may be targets of attack. Low-occupancy (less than 500 people) above- and belowground infrastructure commonly require only two exits according to the International Building Code (IBC, 2007) and therefore have limited emergency access and egress. Choke points may be created when emergency responders move down and occupants move up the same paths.[12] A coordinated terrorist attack may include plans to make exits impassable, creating a greater problem than for surface buildings with windows and direct access to fresh air. More information on terrorism for emergency responders is available in several government sources (see, e.g., FEMA, 2004).

Communication

Surface radio often uses radio repeaters to cover large areas through open air, a technology that may not work in the underground. However, emergency responders critically rely on radio communication. When unable to use surface radio communications technology, responders rely on other technologies including radio repeaters and leaky feeder coaxial cable that functions as extended antennae. These methods work underground, but must be coordinated and robust enough to ensure intelligible coverage throughout the underground through, for example, system redundancy. Interoperability among multiple responders, potentially from many agencies, and the ability to communicate on multiple frequencies are also important to ensure safety. The technology to communicate between emergency responder departments with redundant systems exists today, but these systems may not function in the underground. As mentioned in the previous section, the mining industry is researching enhanced underground communication (e.g., underground use of wireless technologies) between those occupying the underground, and between those on the surface and those underground. Con-

[11] A listing of current and past projects supported in this area of research can be found at http://www.cdc.gov/niosh/mining/researchprogram/communicationstracking.html (accessed October 25, 2012).

[12] The opposite of this happened at the World Trade Center on 9/11.

tinued research in these areas is needed to ensure reliable communications in underground infrastructure.

INCREASING COMFORT AND MAXIMIZING SAFETY

Anecdotal evidence suggests that many people, especially those unaccustomed to being underground on a regular basis, are uncomfortable with the idea of being underground. The committee is not aware of data that quantify the extent of negative perceptions. Negative attitudes may stem from safety concerns, unpleasant personal experiences, or a belief that the underground is dank and dangerous, rather than from specific knowledge about the benefits, risks, and relative safety of underground facilities. Presumably, the public is not campaigning for removal of existing underground systems and services, but it may seem unenthusiastic about new underground applications, especially given their initial costs.[13] Finding ways to clarify and counterbalance negative perceptions can be as big an obstacle as the most complex safety and technical challenges and requires a thorough research focus of its own. The urban underground environment can be engineered and managed—given good design—to be safe, attractive, stimulating, productive, and healthy (Carmody and Sterling, 1993; Meijenfeldt and Geluk, 2003). Given appropriate attention to lighting, ventilation, visual cues for orienting, fire prevention and other safety considerations, emergency egress, and aesthetic considerations, underground space can be as enticing as surface space designed for similar use. Creating underground space that enables and encourages safe, economical, and sustainable use over the long-term is fundamental to that space being part of sustainable and resilient development in the urban setting.

REFERENCES

ASME (American Society of Mechanical Engineers). 2010a. Safety Code for Elevators and Escalators. A17.1/CSA B44 – 2010. New York: ASME.

ASME. 2010b. Guide for Inspection of Elevators, Escalators and Moving Walks. A17.2 – 2010. New York: ASME.

Barrie, A. 2000. War underground: The tunnellers of the Great War. Staplehurst, Kent, United Kingdom: Spellmount Limited.

BRPBTS (The Blue Ribbon Panel on Bridge and Tunnel Security). 2003. Recommendations for Bridge and Tunnel Security. Washington, DC: American Association of State Highway and Transportation Officials (AASHTO) and Federal Highway Administration (FHWA) [online]. Available: http://www.fhwa.dot.gov/bridge/security/brp.pdf (accessed June 14, 2011).

Bugliarello, G. 2001. Rethinking urbanization. The Bridge 31(1):5-12.

Bugliarello, G. 2009. A note on the interfacial vulnerabilities of transportation systems. Pp. 95-103 in Countering Terrorism Biological Agents, Transportation Networks, and Energy Systems: Summary of a U.S.-Russian Network, G.E. Schweitzer, rapporteur. Washington, DC: The National Academies Press.

[13]For example, see Parsons, 2011.

Carmody, J., and R. Sterling. 1993. Underground Space Design: A Guide to Subsurface Utilization and Design for People in Underground Spaces. New York: Van Nostrand Reinhold.

Chapanis, A. 1991. To communicate the human factors message, you have to know what the message is and how to communicate it. Human Factors Society Bulletin. 34(11):1-4.

DiNapoli, T.P., and K.B. Bleiwas. 2010. Financial Outlook for the Metropolitan Transportation Authority. Report 5-2011. New York: Office of the NY State Comptroller [online]. Available: http://www.osc.state.ny.us/reports/mta/mta-rpt-5201l.pdf (accessed April 27, 2011).

DOD (U.S. Department of Defense). 1988. Manpower, Personnel, Training, and Safety (MPTS) in the Defense Acquisition Process. DOD Directive Number 5000.53 (cancelled February 23, 1991, by DOD 5000.1 and DOD 5000.2).

ECTP (European Construction Technology Platform). 2005. Strategic Research Agenda for the European Underground Construction Sector, Draft, October 2005. European Construction Technology Platform [online]. Available: http://www.ectp.org/documentation/FA-UndergroundConstructions_SRA-VISION2030-23Nov05.pdf (accessed May 2, 2012).

Ernst, S., M. Patel, H. Capers, D. Dwyer, C. Hawkins, G.S. Jakovich, W. Lupton, T. Margro, M.L. Ralls, J. Rohena, and M. Swanson. 2006. Underground Transportation Systems in Europe: Safety, Operations, and Emergency Response. FHWA-PL-06-016. Federal Highway Administration [online]. Available: http://international.fhwa.dot.gov/uts/uts.pdf (accessed June 16, 2011).

European Parliament and Council. 2004. Directive 2004/54/EC of the European Parliament and of the Council of 29 April 2004 on minimum safety requirements for tunnels in the Trans-European Road Network. OJEU L167:39-91 [online]. Available: http://eur-lex.europa.eu/LexUriServ/LexUriServ.do?uri=OJ:L:2004:167:0039:0091:EN:PDF (accessed May 29, 2012).

FEMA (Federal Emergency Management Agency). 2004. Fire and Emergency Services Preparedness Guide for the Homeland Security Advisory System, 1st Ed. Emmitsburg, MD: FEMA [online]. Available: http://www.nh.gov/safety/divisions/fstems/ems/preparedness/documents/hsasguide.pdf (accessed May 29, 2012).

Fowler, C. 2008. The Svalbard Global Seed Vault: Securing the Future of Agriculture. The Global Crop Diversity Trust [online]. Available: http://www.croptrust.org/documents/Svalbard%20opening/New%20EMBARGOED-Global%20Crop%20Diversity%20Trust%20Svalbard%20Paper.pdf (accessed December 4, 2012).

Gui, M.W., and M.C. Chien. 2006. Blast-resistant analysis for a tunnel passing beneath Taipei Shongsan airport-a parametric study. Geotechnical and Geological Engineering. 24(2):227-248.

HC (UK House of Commons). 2006. Report of the Official Account of the Bombings in London on 7th July 2005. London: The Stationery Office [online]. Available: http://news.bbc.co.uk/2/shared/bsp/hi/pdfs/11_05_06_narrative.pdf (accessed April 27, 2011).

IBC (International Building Code). 2007. Section 1019.1 Number of Exits and Continuity [online]. Available: http://publicecodes.citation.com/icod/ibc/2006f2/icod_ibc_2006f2_10_sec019.htm (accessed June 27, 2012).

ITA (Association Internationale des Tunnels). 1998. Fire and life safety for underground facilities: Present status of fire and life safety principles related to underground facilities. ITA Working Group 4. Tunnelling and Underground Space Technology. 13(3):217-269 [online]. Available: http://www.itaaites.org/fileadmin/filemounts/general/pdf/ItaAssociation/ProductAndPublication/WorkingGroupsPublication/WG4/Tust_Vol_13_3_217-269.pdf.

ITA-AITES (Association Internationale des Tunnels-International Tunneling and Underground Space Association). 2011. Working Group 5: Health and Safety in Works [online]. Available: http://www.ita-aites.org/index.php?id=75&L=0&tx_ttnews%5Btt_news%5D=27&cHash=4588dfc0bfa5ba891c583bacbae5cfda (accessed June 14, 2011).

Jenkins, B.M., and L.N. Gersten. 2001. Protecting Public Surface Transportation against Terrorism and Serious Crime: Continuing Research on Best Security Practices. Mineta Transportation Institute Report 01-07. Mineta Transportation Institute, San Jose State University [online]. Available: http://www.fta.dot.gov/documents/terrorism_final.pdf (accessed May 29, 2012).

Kantowitz, B.H., and R.D. Sorkin. 1983. Human Factors: Understanding People-System Relationships. New York: John Wiley & Sons.

Licht, D.M., D.J. Polzella, and K. Boff. 1989. Human Factors, Ergonomics, and Human Factors Engineering: An Analysis of Definitions. CSERIAC-89-01. Crew System Ergonomics Information Analysis Center, and H.G. Armstrong Aerospace Medical Research Laboratory, Wright Patterson AFB, Dayton, OH [online]. Available: http://www.hfes.org/Web/EducationalResources/HFDefinitions.pdf (accessed June 19, 2011).

Linger, D.A., G.H. Baker, and R.C. Little. 2002. Applications of underground structures for the physical protection of critical infrastructure. Pp. 333-342 in North American Tunneling, L. Ozdemir, ed. Lisse, The Netherlands: Swets and Zeitlinger [online]. Available: works.bepress.com/cgi/viewcontent.cgi?article=1002&context=george_h_baker&sei-redir=1#search="Underground+Infrastructure+Security (accessed April 26, 2011).

McCamley, N.J. 1998. Secret Underground Cities: An Account of Some of Britain's Subterranean Defence, Factory and Storage Sites in the Second World War. South Yorkshire, United Kingdom: Pen & Sword Books, Ltd.

Meijenfeldt, E.V., and M. Geluk. 2003. Below Ground Level: Creating New Spaces for Contemporary Architecture. Basel, Boston, Berlin: Birkhäuser.

Miles, S.B. 2011. The role of critical infrastructure in community resilience to disasters. Pp. 1985-1995 in Proceedings of the 2011 Structure Congress, April 14-16, 2011, Las Vegas, NV, D. Ames, T.L. Droessler, and M. Holt, eds. Reston, VA: ASCE. DOI:10.1061/41171(401)173.

Molag, M., and I.J.M. Trijssenaar-Buhre. 2006. Risk Assessment Guidelines for Tunnels. Safe and Reliable Tunnels, Second International Symposium, Lausanne, Switzerland [online]. Available: http://www.ita-aites.org/fileadmin/filemounts/general/pdf/ItaAssociation/Committees/COSUF/Molag.pdf.

Myers, C.W., and N.Z. Elkins. 2009. Underground Collocation of Nuclear Power Reactors and Repository to Facilitate the Post-Renaissance Expansion of Nuclear Power. Waste Management Conference, March 1-5, 2009. Phoenix, Arizona [online]. Available: http://permalink.lanl.gov/object/tr?what=info:lanl-repo/lareport/LA-UR-08-07452 (accessed December 4, 2012).

NFPA (National Fire Protection Association). 2010a. NFPA 130: Standard for Fixed Guideway Transit and Passenger Rail Systems [online]. Available: http://www.nfpa.org/aboutthecodes/AboutTheCodes.asp?DocNum=130&cookie%5Ftest=1 (accessed June 9, 2011).

NFPA. 2010b. NFPA 520: Standard on Subterranean Spaces [online]. Available: http://www.nfpa.org/aboutthecodes/AboutTheCodes.asp?DocNum=520 (accessed June 9, 2011).

NFPA. 2010c. NFPA 1710: Standard for the Organization and Deployment of Fire Suppression Operations, Emergency Medical Operations, and Special Operations to the Public by Career Fire Departments [online]. Available: http://www.nfpa.org/aboutthecodes/AboutTheCodes.asp?DocNum=1710 (accessed June 8, 2012).

NFPA. 2011. NFPA 502: Standard for Road Tunnels, Bridges, and Other Limited Access Highways [online]. Available: http://www.nfpa.org/aboutthecodes/AboutTheCodes.asp?DocNum=502 (accessed June 9, 2011).

NRC (National Research Council). 2007. Human-System Integration in the System Development Process: A New Look. Washington, DC: National Academies Press.

OECD (Organisation for Economic Co-operation and Development). 2006. OECD Studies in Risk Management: Norway Tunnel Safety. Paris, France: OECD [online]. Available http://www.oecd.org/dataoecd/36/15/36100776.pdf (accessed June 19, 2011).

OSHA (Occupational Safety and Health Administration). 2003. Underground Construction (Tunneling). OSHA 3115-06R 2003. Washington, DC: OSHA [online]. Available: http://www.osha.gov/Publications/osha3115.pdf (accessed June 8, 2012).

Parsons, J. 2011. Virginia Pol Criticized Dulles Station Design, Labor Agreements. ENR, April 19, 2011 [online]. Available: http://enr.construction.com/infrastructure/transportation/2011/0419-DullesStationCriticized.asp (accessed April 20, 2011).

Prager, F., G. B. Asay, B. Lee, and D. von Winterfeldt. 2010a. Exploring reductions in London Underground passenger journeys following the July 2005 bombings. In Estimating Behavioral Changes for Transportation Modes After Terrorist Attacks in London, Madrid, and Tokyo, Final Report, D. Von Winterfeldt and F. Prager, eds. METRANS Project 08-10. National Center for Risk and Economic Analysis of Terrorism Events [online]. Available http://www.metrans.org/research/final/08-10%20Final%20Report.pdf (accessed October 23, 2012).

Prager, F., B. Fasolo, Z. Ni. 2010b. Analysis of passengers' reactions to the sari gas attacks in Tokyo. In Estimating Behavioral Changes for Transportation Modes After Terrorist Attacks in London, Madrid, and Tokyo, Final Report, D. Von Winterfeldt and F. Prager, eds. METRANS Project 08-10. National Center for Risk and Economic Analysis of Terrorism Events [online]. Available http://www.metrans.org/research/final/08-10%20Final%20Report.pdf (accessed October 23, 2012).

Rogoza, J., and P. Zochowski. 2010. Attacks in the Moscow Metro. East Week 13(206) [online]. Available: http://www.osw.waw.pl/en/publikacje/eastweek/2010-03-31/attacks-moscow-metro (accessed April 27, 2011).

Rowshan, S., W.C. Sauntry, T.M. Wood, B. Churchill, and S.R. Levine. 2005. A Guide to Physical Security Risk Management for Transportation Management Centers. Washington, DC: Federal Highway Administration [online]. Available http://trb.metapress.com/content/37pv8315628j5724/fulltext.pdf (accessed June 8, 2012).

SAIC (Science Applications International Corporation). 2002. A Guide to Highway Vulnerability Assessment for Critical Asset Identification and Protection: Contractor's Final Report. Prepared for the American Association of State Highway and Transportation Officials' Security Task Force, by Science Applications International Corporation, Vienna, VA [online]. Available: http://security.transportation.org/sites/security/docs/guide-VA_FinalReport.pdf (accessed June 14, 2011).

TRB (Transportation Research Board). 2006. Transportation Security, Volume 12: Making Transportation Tunnels Safe and Secure. Transit Cooperative Research Program Report 86/National Cooperative Highway Research Program Report 525. Washington, DC: TRB.

Tu, A.T. 1999. Overview of sarin terrorist attacks in Japan. Pp. 304-317 in Natural and Selected Synthetic Toxins, A.T. Tu and W. Gaffield, eds. ACS Symposium Series 745 [online]. Available: http://pubs.acs.org/doi/abs/10.1021/bk-2000-0745.ch020 (accessed April 27, 2011).

Von Winterfeldt, D., and F. Prager. 2010. Estimating behavioral changes for transportation modes after terrorist attacks in London, Madrid, and Tokyo, Final Report. METRANS Project 08-10. National Center for Risk and Economic Analysis of Terrorism Events [online]. Available http://www.metrans.org/research/final/08-10%20Final%20Report.pdf (accessed October 23, 2012).

UNECE (United Nations Economic Commission for Europe). 2001. Recommendations of the Group of Experts on Safety in Road Tunnels Final Report. TRANS/AC.7/9 [online]. Available: http://live.unece.org/fileadmin/DAM/trans/doc/2002/ac7/TRANS-AC7-09e.pdf (accessed June 22, 2011).

U.S. Department of State. 2004. Patterns of Global Terrorism, 2003 [online]. Available: http://www.state.gov/documents/organization/31912.pdf (accessed June 14, 2011).

WMATA (Washington Metropolitan Area Transit Authority) and Cambridge Systematics, Inc. 2009. Transit Ridership Trends and Markets, March 2009 [online]. Available: http://www.wmata.com/pdfs/planning/FINAL%20Transit%20Ridership%20and%20Market%20Trends%20Report.pdf (accessed April 27, 2011).

5

Lifecycle Sustainability, Costs, and Benefits of Underground Infrastructure Development

Underground development provides opportunities to use available urban space more effectively, but it requires significant and potentially cost-prohibitive investment for initial construction as compared to similar-use infrastructure built on the surface. This chapter summarizes the existing knowledge about the lifecycle sustainability, costs, and benefits of underground development.

Literature concerning impacts of underground infrastructure on the lifecycle sustainability of urban development is relatively scant. More is known about monetary lifecycle costs and benefits, while less is known about long-term environmental or social impacts. Even those studies related to economic benefits and costs were primarily to inform assessment of alternatives for proposed projects, such as for the Alaska Way Viaduct in Seattle (Taylor, 2008). Fewer retrospective studies have been conducted to assess actual costs and benefits of underground development.

This chapter does not provide a lifecycle cost assessment for any underground works; rather it identifies factors to be considered in a lifecycle assessment in terms of economic costs and benefits throughout the infrastructure life (construction, operation, and renovation) and environmental and social costs and benefits. Research that would inform better and more comprehensive lifecycle assessments is identified.

LIFECYCLE SUSTAINABILITY ASSESSMENT

In assessing lifecycle sustainability, a "triple bottom line" analysis is often adopted that considers the economic, environmental, and social impacts of development. Elkington introduced the basic concepts of the approach in 1994 and expanded on them and introduced the term "triple bottom line" in 1997 (Elkington, 1994, 1997). The approach provides a framework for a multiple objective assessment of complex investments. "Full cost accounting" pursues a similar goal of including a wide range of impacts in decision making, but full cost accounting usually focuses on developing monetary estimates of different impacts. A recent example of this approach was the estimate of external costs associated with energy production (NRC, 2010). However, environmental and social impacts are difficult to quantify monetarily and often are beyond the current state of knowledge about underground development because of lack of attention. Accordingly, this chapter is divided into sections that consider the lifecycle economic, environmental, and social impacts of underground development. This review of lifecycle costs and benefits is consistent with the committee's task to explore how use of the underground could increase sustainability.

Lifecycle Planning and Assessment

Underground development often involves a relatively long life cycle even when compared with other infrastructure investments. For example, the Circle Line subway in London was originally constructed more than 150 years ago in the mid-nineteenth century (Bobrick, 1981). Although the line has been extended, renovated, and rehabilitated over time, the original investment in underground construction is still paying off and providing travel and other benefits.[1] Similarly, underground pipelines can also last for more than 100 years, especially if in situ inspection, cathodic protection,[2] and rehabilitation are performed (e.g., MWRA, 2006). However, government and private planning horizons are usually fairly short with respect to the useful life of the infrastructure. Metropolitan and state-wide long-range transportation plans, for example, often consider the benefits and costs of investment for only a 20-year horizon (DOT, 2007). Such a short planning horizon means that any benefits from underground development that occur after 20 years are not considered in investment decision making.

Underground infrastructure development involves an initial investment to create usable space that provides benefits over an extended period. Long lifetimes of underground infrastructure may be excluded from analyses performed by those with short planning horizons, just as owner and user costs of renovating surface facilities may be excluded from cost analyses, although they may be quite

[1] For example, to provide shelter. London subway tunnels were used as bomb shelters during World War II.

[2] Corrosion protection.

large. Similarly, high discount rates for calculating the present value of future benefits will make those benefits less valuable if provided long into the future. For example, the federal fiscal year 2011 real test discount rate for a 30-year planning horizon was established at 2.7 percent (OMB, 2010). With this discount rate, \$1.00 of real benefits received 30 years later would have a 2011 value of $1/1.027^{30}$ = \$0.45, or less than half. One dollar of benefits received 100 years in the future would either be disregarded as beyond the planning horizon or would have a 2011 value of only \$0.07.

A long lifetime in itself also may affect planning for future alternatives. Particular underground development can preclude other uses or make them more expensive to implement. For example, underground transportation tunnels such as the Boston Central Artery project required rebuilding and relocating existing underground utilities in the tunnel right-of-way. Building foundations may make re-use of their underground locations prohibitively expensive, precluding new underground parking, tunnels, or other uses in that location. In effect, underground construction may increase cost and reduce flexibility of options for alternative future uses. Because most underground facilities are left in the ground even after their useful life ends, the extra cost or difficulty of re-using the space continues nearly indefinitely. A comprehensive planning effort would recognize that underground space is a resource that should be used in the best manner possible, rather than letting initial uses preclude later uses. Similar conclusions have been drawn with respect to limiting space debris in orbits around Earth that may prevent use of those orbits for other purposes (e.g., UN, 1999).

In addition to assessing the life cycle of underground infrastructure itself, sustainability suggests that impacts of the infrastructure also be considered for the entire life cycle of a project. Lifecycle assessment "studies the environmental aspects and potential impacts throughout a product's life (i.e., cradle-to-grave) from raw material acquisition through production use and disposal" (ISO, 1997). Figure 5.1 illustrates a generic supply chain life cycle. For underground infrastructure, the supply chain would include the various materials and processes involved in construction as well as inputs such as energy for lighting and ventilation during facility operation. Closure and decommissioning costs would be included in the disposal phase in Figure 5.1. The landfill phase would be expected to include the costs of providing an engineered landfill for disposal or any costs associated with legacy structures underground.

Metrics to use in assessing sustainable development overall, as well as to assess specific economic, environmental, and social impacts, are still a subject of widespread debate even without consideration of the special circumstances of underground development (Jeon and Amekudzi, 2005). Economic impacts are typically expressed in monetary units, but a variety of impacts may be considered for environmental and social impacts. For example, Reijnders (1996) suggests that broad environmental impacts be considered in preparing a lifecycle assessment including:

- impact on resources (e.g., use of renewable and nonrenewable resources, pollution of resources);
- direct impact on nature and landscape, such as through undesirable change in landscape;
- air pollution and its contribution to climate change, smog, acid deposition, odors, and deterioration of the ozone layer;
- soil pollution, such as solid wastes added to soil, through eutrophication, added toxins, and contributions to groundwater pollution;
- surface water impacts, including biological or chemical discharges with oxygen demand, toxic discharges, surface water warming, and contribution to eutrophication;
- noise;
- electromagnetic radiation or fields; and
- ionizing radiation.

In many environmental lifecycle assessment studies, environmental impact estimates are limited to only a few critical categories of impacts, such as emissions of greenhouse gases and conventional pollutants.

Assessing system interdependencies over the life cycle of underground infrastructure is also an important and challenging part of assessing risk. A variety of analytic tools exist to aid in risk assessment of individual infrastructure systems and interactions. These include Bayesian networks, Monte Carlo simulation, and decision trees (Rinaldi et al., 2001; Haimes, 2004; Weber et al., 2012). Applying such tools may inform decision making by reducing some of the high levels of uncertainty associated with different kinds of risk, especially when dealing with interactions of complex systems.

In this chapter, the committee assembles existing knowledge of the impacts of underground development, recognizing there are numerous knowledge gaps, especially because past studies generally took a narrower view of benefits and costs than is required for a lifecycle sustainability perspective. There is also considerable variation and uncertainty in the performance of underground development, especially with regard to extreme events such as earthquakes or flooding. Moreover, the general advantages and disadvantages for underground facilities described in Chapter 3 necessitate specific evaluations for each type of use and site circumstances.

LIFECYCLE ECONOMIC BENEFITS AND COSTS

Increasing population, consumption, density, globalization, communication, and other trends suggest an increasing complexity for human society (Boyle et al., 2010). Implementation of technological advancements can have both positive and negative repercussions. It is important that processes to deliver sustainable

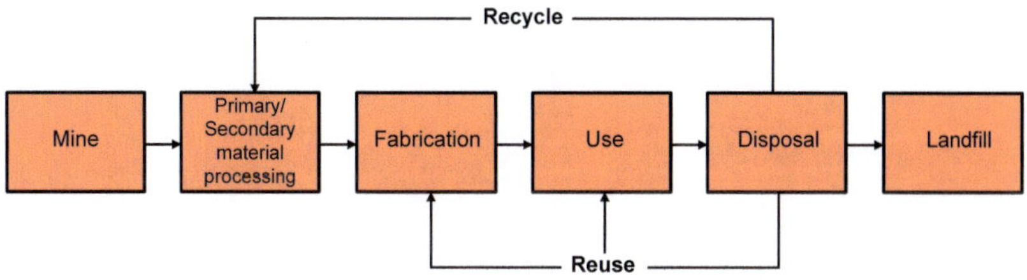

FIGURE 5.1 A generic product cradle-to-grave life cycle. SOURCE: Modified from Hendrickson et al., 2006.

underground infrastructure be carefully designed to limit negative impacts while gaining the maximum benefit. Indirect impacts of technology advancements also must be considered. Developing underground space provides the opportunity to use surface space for other purposes such as green space for parks or other aboveground development within or closer to urban centers, but quantification of such opportunities is difficult. Low-impact design of infrastructure systems that reduces environmental impacts and transportation costs is now being incorporated into urban development (TRB, 2009). Compact city trends support an underground development concept including a wide range of underground facilities that contribute to an efficient but highly livable environment. In this regard, inherent economic benefits are derived from utilizing the subsurface as part of the provision of housing, transportation, commercial, industrial, and utility facilities.

Intensive Development and the Compact City

There has been a longstanding debate about the benefits and costs of intensive development in the form of compact cities relative to dispersed development and urban sprawl (e.g., Ewing, 1997; Gordon and Richardson, 1997). Compact cities are distinguished by high densities of people per unit land area, a mix of land uses within neighborhoods, one or more high-density centers of employment, and careful spatial arrangement or contiguity of land uses (NRC, 2010). Critics of the compact city note the deleterious effects of more intensive development, including increased traffic congestion, less affordable housing, and higher consumer costs (Gordon and Richardson, 1997; O'Toole, 2009).

A recent NRC study found that compact cities are likely to reduce vehicle miles of travel and both direct and indirect energy consumption and greenhouse gas emissions (NRC, 2010). European experience is similar (Schwanen et al., 2004). Shammin and others (2010) estimated that total energy use is roughly 17 percent lower for urban area residents than for rural or low-density area residents, even when all purchased goods and services are considered. To some extent, these expected benefits from compact cities may arise from self-choice of residents who wish to drive less, but even when attitudinal factors are taken into account,

FIGURE 5.2 Boston Central Artery as an elevated structure and as an underground roadway SOURCE: MADOT, 2012.

less vehicle miles are traveled in compact cities (Handy et al., 2005). Compact development also may reduce infrastructure costs and development pressure on green spaces (Ewing, 1997).

However, higher urban density seems to directly correlate with higher levels of underground space development (Sterling et al., 2012). Many planners believe that underground development and use could enhance the net benefits of intensive development. Use of underground space can reduce traffic congestion and the consumer costs noted by critics of compact cities while simultaneously achieving the travel and energy reductions identified by compact city proponents. Figure 5.2 shows the Boston Central Artery, which was originally built as an elevated structure through downtown Boston but was moved underground in the Big Dig project, resulting in a corridor of open space (NAE/NRC, 2003). The net benefits and indirect effects on long-term development may be significant even though they are difficult to assess on a project-by-project basis.

Construction Phase Economic Benefits and Costs

Our current "built environment includes buildings, engineering works, and infrastructure such as roads, wastewater and water treatment plants, storm water management systems, power generation facilities, railways, bridges, and even natural systems such as rivers and harbors" (Boyle et al., 2010). Underground development provides an opportunity to place many of these facilities in the

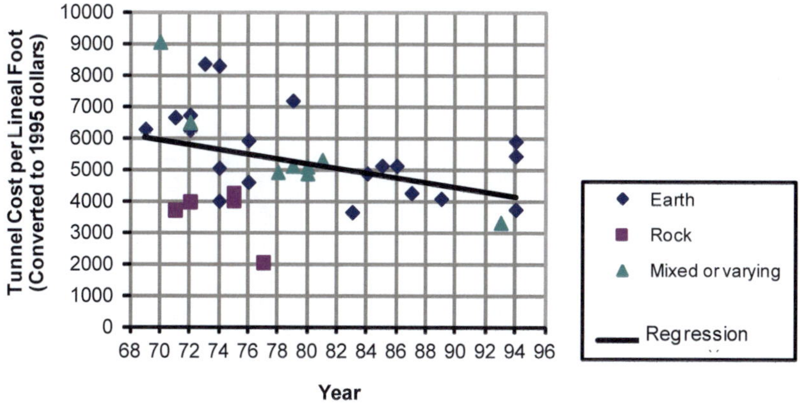

FIGURE 5.3 Cost for mining and lining approximately 20 ft. diameter tunnels for the Washington Metro over the period 1969-1994. SOURCE: R. Sterling, from data supplied by WMATA (courtesy of Walt Mergelsberg). Reprinted with permission of author.

largely available space—real estate—beneath existing surface developments. Sterling (2005) describes the importance of urban underground space planning. Initial costs for underground construction include those related to geological site characterization and management of geologic conditions, finding and relocating utilities, potential disruption to existing infrastructure due to utility strikes, requirements for engineered backfill, and traffic control along a horizontal alignment. A nationwide effort exists to use best practices in underground works in the interest of public safety (CGA, 2008). However, in urban areas, existing structures constrain practical design of underground facilities. Underground facilities must accommodate facility design restrictions and land or easement availability for construction. The time associated with accommodating requirements associated with environmental and safety regulations also must be factored into construction costs.

Figure 5.3, based on data from construction of the Washington Metro from 1969 to 1994, shows a decreasing trend line for raw tunnel construction costs and, equally importantly, a narrowing of the costs range over this 25-year period. Although project costs are highly dependent on specific circumstances, for example, the difficulty of installation of specific sections, this graph could suggest that accumulated knowledge and risk management, investments in research, and adoption of better technologies and contracting practices over the period resulted in cost reductions for actual tunnel construction. Are these cost reductions being seen in the total cost of newer underground construction projects? The answer is probably "no," because demands for higher safety standards and reduced construction risk and environmental impact for newer projects have increased. In addition, such changes in technical costs may be

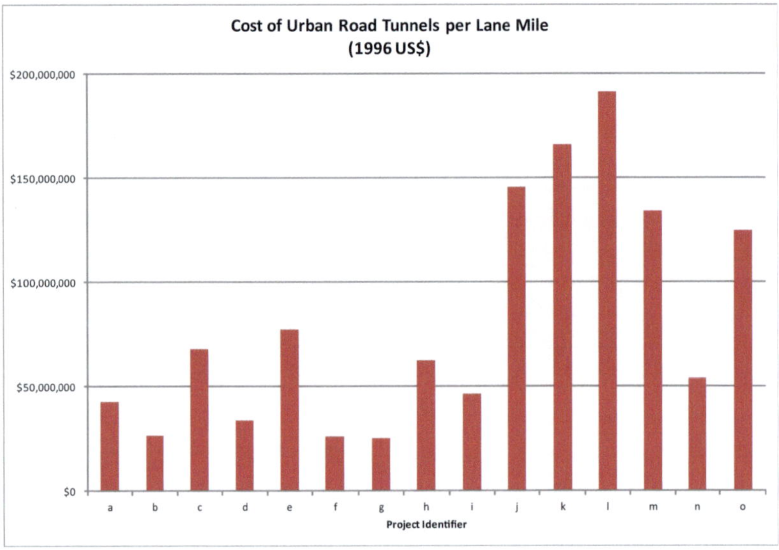

FIGURE 5.4 Variability in urban road tunnel costs based on data from Australia, France, Japan, Sweden, and the United States. Each letter on the x axis represents a different project for which cost data were provided. SOURCE: R. Sterling, from data collected by ITA Working Group 13. Modified with permission of the author.

masked by the wide range in total project costs seen in worldwide projects. For illustration, Figure 5.4 depicts data collected during a study of the costs and benefits of underground transportation facilities undertaken by the International Tunneling and Underground Space Association (ITA) Working Group 13 (ITA WG13, 2004). Cost data on road tunnels from 30 cities in 19 countries were compiled from questionnaire responses and were converted to a common basis in terms of 1996 U.S. dollars for the cost per lane mile of roadway.

Although these costs vary because of what is counted in each project's costs, the variations observed among countries in constructing a lane kilometer of roadway suggest that it would be worthwhile to investigate the reasons for lower costs in some countries as compared to others. Although local geology may have a role, it is not expected to be the only significant reason for the observed variations. Differences in design standards, administrative review processes, public engagement, and streamlining of design and construction processes may be increasingly important. Understanding the reasons for varying costs is important so that the outcomes of projects developed in other countries can be judged according to standards other than high cost and so that factors contributing to high costs can be identified and improved.

Various estimates of the length of water, sewer, and storm water pipelines in the United States can be found in the literature. Table 2.1 lists a total of approximately 3.7 million miles of pipeline including transmission, distribution, and private service connections. More than 480,000 km of underground utilities are estimated to be installed worldwide annually, including water, sewer, gas, electrical, cable television, and telephone (Najafi, 2005). A significant portion of this infrastructure is buried beneath paved surfaces in urban environments. Consequently, more efficient and effective installation and rehabilitation of this vast utility network would provide significant economic benefits due to lower direct cost and a minimal disruption of this surface environment.

Lane closures due to surface construction and the subsequent detours cause traffic delays and have an impact on the cost of fuel (CNRC, 2005). Impacts can be minimized through the selection of suitable construction equipment. Further savings for initial capital equipment may be realized, for example, with trenchless methods, especially in horizontal construction because of reduced use of construction equipment (Woodroffe and Ariaratnam, 2008). In contrast, open-cut excavation requires the use of numerous pieces of equipment including excavators, bulldozers, surface compactors, and haul vehicles.

Now implemented in underground works are alternative contracting mechanisms that provide innovative means for allocating project risks to reduce their effects on bid amounts. These include approaches such as design-build, design-build-own-operate-transfer, and construction manager at risk. Additionally, performance-based specifications are used to promote contractor creativity and reduce construction costs. Incompleteness of performance-based specifications, however, may negatively affect the final product.

There is little comparison of the costs of underground versus aboveground construction (Parker, 2007). Lifecycle cost analyses consider the direct, social, and environmental costs as well as the costs for specialty items such as heating, ventilation, and air conditioning systems over the life cycle. Because they are critical to infrastructure functionality and must be carefully selected and installed during initial construction, these and other operational costs usually are combined with direct capital costs in selecting the best construction alternative.

Safety hazards and risks are inherent in all construction projects and need to be assessed during the design phase. A risk-based safety impact assessment approach was adopted for the construction of a subway line in Seoul, Korea (Seo and Choi, 2008). Open-cut construction also was evaluated for comparison purposes. The goal was to identify and reduce, prior to construction, the risks associated with design items that could cause construction accidents. This is important because subsurface construction is done "out of sight," thereby requiring a high degree of skill and extensive experience on the parts of the designer (often contractually obligated to provide full-time quality control inspection) and the constructor. The design and construction of subsurface infrastructure represent unique scenarios in which design, inspection, and construction functions cannot easily be separated (Kagan et al., 1986).

Operation Phase Lifecycle Economic Benefits and Costs

As noted earlier, cost benefits accrued from operation of any infrastructure system are difficult to quantify. Benefits may include enhancements to quality of life, reductions in travel and travel time, and increases in productivity. There are, however, inherent benefits related to the operation of underground infrastructure. Johnson (2006) found that the conversion of unsightly overhead electrical lines to underground lines resulted in increased property values and improved aesthetics within neighborhoods. Other lifecycle societal economic benefits include reduced outages, transmission losses, and greenhouse gases; reduced network maintenance costs; fewer electrocutions; and fewer motor vehicle collisions with poles (IFC Consulting, 2003). The average cost of burying existing electrical lines is estimated to be $1 million per mile, which is almost 5 to 6 times (Parsons Brinkerhoff, 2012) or 10 times (Johnson, 2006) the cost of a new overhead line. However, the maintenance and operating costs of underground electrical lines have been reported to be about one-tenth of those of aboveground lines because of reduced transmission losses over the life cycle (IFC Consulting, 2003). In addition, underground cables also may enable increases in power transmission capacity (Al-khalidi and Kalam, 2006).

Chapter 4 describes security issues associated with underground infrastructure but shows that there are inherent security benefits to putting infrastructure underground. Underground systems have a lower risk of disaster failure to earthquakes, hurricanes, tornados, tropical storms, heavy snow events, monsoon winds, and natural disasters, but these systems may be vulnerable to flooding. These lower risks could translate into reduced insurance premiums over the life cycle of the asset (e.g., De Saventhem, 1977).

Renovation and Replacement Phase Lifecycle Economic Benefits and Costs

Renovation of infrastructure (i.e., asset preservation) often improves operation at a fraction of the cost of full replacement. Consequently, renovation methods such as lining or grouting of pipelines and external face-lifting of buildings are preferred when existing infrastructure is still structurally acceptable but requires renewal to a "like new" condition. Replacement may be deemed necessary because of obsolescence, inflexible design, or irreparability of the existing infrastructure. Surface infrastructure can be replaced with relative ease as compared to underground infrastructure; however, the frequency of the need for repairs and renovations may be less for underground infrastructure because of the protection the underground provides. On the other hand, if underground infrastructure becomes obsolete—for example, the largely abandoned underground freight tunnel system beneath downtown Chicago (see Box 3.7)—it may be difficult to repurpose the space for another use.

Careful planning of underground use for well into the future can minimize the rate at which infrastructure becomes obsolete. Utilidors (described in Chapter 3), for example, provide flexibility to switch out or add utilities when dictated by obsolescence, deterioration, or capacity issues. Utilidors streamline utility easements and provide improved accuracy in locating existing buried utilities, which is advantageous for line maintenance and replacement. Canto-Perello and others (2009) found that utilidors minimize the potential dilemma of mutual interference between utilities and transportation networks. Additionally, placing utilities in utilidors results in minimizing physical damage to surface streets from continual cutting of pavement when installing, inspecting, maintaining, repairing, or replacing lines.

LIFECYCLE ENVIRONMENTAL BENEFITS AND COSTS

Since 1970, the National Environmental Protection Act (NEPA) has required "federal agencies to integrate environmental values into their decision making processes by considering the environmental impacts of their proposed actions and reasonable alternatives to those actions" (EPA, 2010). As a result, environmental impact statements and analyses have been completed for a wide range of underground developments. However, these impact statements are prospective in nature to inform planning decisions, rather than retrospective assessments of actual environmental impacts from projects as built. For example, whereas many earlier environmental impact analyses did not include greenhouse gas emission effects, recent environmental impact statements address findings such as the 2009 finding by the EPA Administrator that greenhouse gas emissions (carbon dioxide, methane, nitrous oxide, hydrofluorocarbons, perfluorocarbons, and sulfur hexafluoride) threaten public health and the welfare of current and future generations as a result of climate change effects (EPA, 2009). Finally, environmental impact statements typically do not include the supply chain or indirect environmental impacts in the analyses and therefore do not provide a complete lifecycle assessment. For estimating carbon footprint or greenhouse gas emissions, these indirect emissions are termed Tier 3 emissions and often are significant for the provision of goods and services (Matthews et al., 2008). In particular, the production of cement used in underground construction generally results in significant greenhouse gas emissions.

Construction methods play a major role in greenhouse gas emissions. Sihabuddin and Ariaratnam (2009) compared airborne emissions from trenchless versus open-cut pipe replacement on the same project and found that trenchless reduced pollution on the order of 80 percent. Few studies have looked at the effect of underground infrastructure over its entire life cycle or have compared lifecycle assessments of overhead and underground infrastructure delivering the same service (see Box 5.1).

BOX 5.1
Environmental Lifecycle Comparison of Overhead and Underground Power Distribution

Bumby et al. (2010) compared buried and overhead power distribution using Southern California Edison designs for medium voltage cables using a process-based lifecycle assessment per guidelines from the International Organization for Standardization.[a] The Figure shows the various process steps involved in the life cycle for the underground power distribution assembly. Their assessment indicates that overhead distribution assemblies as designed by Southern California Edison have lower overall emissions. The values are heavily influenced by the additional material inputs required for cable manufacturing of the underground distribution assembly. Secondary factors include the shorter estimated life for underground cables due to underground heating effects and lost carbon sequestration due to timber production because carbon is captured in the growth of trees. The study also estimated eco-indicator impacts common in Europe (see Guinée, 2002, for standards), including abiotic depletion potential, acidification potential, eutrophication potential, freshwater aquatic ecotoxicity, human toxicity potential, photochemical ozone creation potential, and terrestrial ecotoxicity potential. For reasons similar to those for greenhouse gas emissions, the overhead design had lower environmental impacts in these categories.

The study omitted some categories that require further research. The underground cable had lower resistance, so transmission power losses may be lower underground. The study does not consider land use impacts and the net urban system energy usage or environmental effects given either overhead or underground use. The construction material advantage for power cables may not exist for overhead structures used for other purposes such as carrying vehicles. Moreover, siting overhead power transmission lines often can be difficult for aesthetic reasons. This study demonstrates the difficulty of obtaining comprehensive but rigorous results from triple bottom line analyses. Such analyses can include only the issues for which data are available and are unable to address broader performance, resilience, societal, or environmental issues.

SOCIAL BENEFITS AND COSTS

This section summarizes some of the social benefit and costs associated with the use of underground space and discusses what additional data or changes in assessment practices might be helpful to making sound investment and operational decisions.

As described in earlier sections, the framework for the economic and environmental lifecycle assessment of project alternatives is reasonably well understood—including how to manage conceptually the combination of quantitative and subjective comparisons. One challenge to lifecycle cost analysis is that some of the strongest advantages of underground structures tend to be more long term and qualitative (including benefits to quality of life or urban resilience), while disadvantages tend to be more readily identifiable and quantifiable (e.g., startup

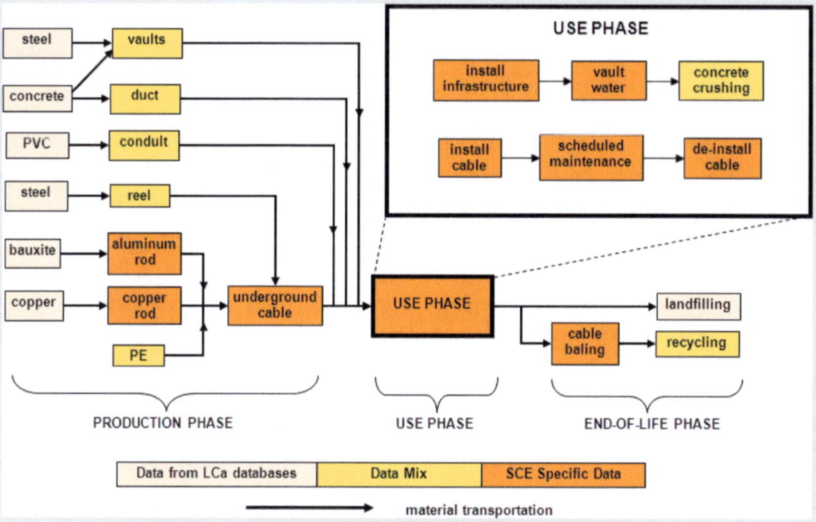

FIGURE Process flow diagram for the underground power distribution assembly. Colors indicate data source as commercial lifecycle assessment databases (pink), Southern California Edison (orange), or a mix of these two sources (yellow). SOURCE: Bumby et al., 2010. Reprinted with permission from American Chemical Society.

[a]ISO 14044 specifies requirements and provides guidelines for lifecycle assessment including scope, inventory analysis phase, impact assessment phase, interpretation phase (ISO, 2006).

costs). Another challenge is that few individuals are expected to state a preference for being in an underground facility rather than a surface facility for extended periods. In many cases, the benefits come from what the underground facility permits in terms of an improved surface environment, mobility, or services rather than from the superior attributes of the facility itself.

Underground space use, if well planned, permits excellent options for urban transportation and provision of utility services, along with a range of other desired facilities, all with low-impact on the surface environment, heritage, and, potentially, ecology. In other words, well-planned underground construction supports a compact, well-functioning, livable, and sustainable urban environment. The protection and resilience of an underground structure may benefit the project owner, but if it affects the ability of society to function effectively, for example, after a disaster, then it has a much broader societal impact. Likewise, communities are increasingly resisting construction-caused disruption from new infrastructure projects. The project owner may pay some costs attributable to the disruption—such as business loss—but the owner does not pay for traffic delay costs and the diminished livability of the neighborhood due to construction noise, vibration, dust, and diminished air quality. Capturing all of the appropriate costs when comparing project alternatives remains a challenge and a topic for future research.

Multiple papers identify issues to be considered with respect to utility projects (e.g., Gilchrist and Allouche, 2005) and provide case examples of the application of social and indirect costs to project decision making (e.g., Li et al., 2009). However, typically only a few of the key social or indirect costs are considered because of a lack of impacts data or a lack of accepted costing for disturbances effects. Papers that describe analyses of a variety of costs (e.g., Pucker et al., 2006) typically find that traffic delay costs are the most important social cost in urban areas and can rival or exceed the cost of the construction itself for some street utility work. In suburban or rural areas, traffic delays are typically less severe except on key arterial routes.

Local opposition to a project typically is based on the social and indirect costs expected as a result of project construction and operation. Often, these costs can be mitigated through less disruptive construction methods (e.g., trenchless technologies for utility construction and repair, and bored tunnels instead of cut-and-cover tunnels for road and rail projects) and restrictions or modifications to working practices (e.g., limits on working hours, noise, and vibration). As restricted working practices are adopted to accommodate neighborhood opposition, unpaid social costs become hard construction costs and potentially increase construction risks. Least disruptive construction methods are more likely chosen, avoiding the need to calculate social costs.

Another issue worth noting is that construction and operation impacts of major infrastructure projects represent a moving target in terms of acceptable compromises for limiting impacts on neighborhoods. Discussions about transforming a surface or elevated transportation project to an underground alignment, or transforming from cut-and-cover to bored tunnel construction, typically consider noise and air quality impacts at the tunnel portal. In general terms, the shift underground maintains mobility for many people in the urban area and lessens the environmental impact on most of the area through which it passes. However, construction vibrations (e.g., from blasting) and noise and air quality emissions become more localized—making them more bothersome to those in the immediate vicinity, but also more controllable. The drawback is that the increasingly high standards to which underground projects may be held increases their costs relative to surface or elevated alternatives. Critical decisions regarding major infrastructure initiatives for urban areas ride on such concerns. The ability to adequately compare radically different infrastructure alternatives (including the "do nothing" alternative) that potentially change the face of the city for better or worse remains a daunting challenge. In many cases, a strong political decision is finally made in the face of widely different opinions and conflicting cost-benefit analyses.

Accommodating social and human factors issues and improving underground designs are not just window dressing essentially technical projects. How these issues are addressed in the project's design and construction can have profound effects on its cost, acceptance by the public, and impact over its life cycle. There is no single best answer, but it is important to understand the various ramifications. The Stockholm (Sweden) Metro has individualized station designs decorated by artists to make distinctly different environments in each station (Winqvist and Mellgren, 1988). Washington, DC, Metro stations have a similar look that creates familiarity for ease of use. Large station caverns often are used to create an impressive public space underground, but at a cost in terms of initial construction and probably in operation as well (as pointed out by O'Rourke, 1983). Allowing variety in design approaches based on a better understanding of how to create interesting and enjoyable underground spaces without large increases in cost or space requirements remains a challenge, as does quantifying the social costs and benefits over the life cycle of the infrastructure.

RESEARCH NEEDS FOR LIFECYCLE COSTS AND BENEFITS

As discussed earlier, many factors are incorporated into full lifecycle cost analysis. Consideration of those factors may shift the perception of the feasibility of underground space use—from that of expensive and risky, to wise and most cost-effective in the long term. Largely needed is a better understanding of what aspects of project planning, design, construction, and operation contribute the most to project costs and long-term benefits and performance. The goals of lifecycle cost analysis are to reduce costs where possible through technology enhancements and design and administrative changes, as well as to better articulate the long-term benefits to the urban area—in monetary terms if possible—but, at least through well-documented examples of the positive and negative impacts of underground projects.

Considering the high profile of many underground road and rail projects, it is surprising that comprehensive documentation is hard to find. Planning studies are available, but they lack the retrospective assessment of actual costs and benefits. There is anecdotal or partial evidence of the positive environmental and financial impacts of replacing aboveground transportation structures with underground alignments on neighborhoods worldwide. For example, the Boston Globe reported in 2004 (Palmer, 2004):

> *According to an in-depth review of the City of Boston tax assessing records by the Globe, in the 15 years since the Central Artery tunnel project began, the value of commercial properties along the mile-long*

strip that this year will become the Rose Kennedy Greenway increased to $2.3 billion, up 79 percent. That's almost double the citywide 41 percent increase in assessed commercial property values in the same period.

When adjusted and aggregated over the entire Central Artery alignment, the increase in land values could be of the same order of magnitude as the cost of such a difficult and expensive project. What appears to be lacking in this and other examples is careful and defensible study of the financial and environment changes over, say, a decade following project completion. Retrospective, comparative studies of the costs and impacts of the various types of underground construction projects are needed. To be useful, these studies must be conducted in a comprehensive and scientific manner and must consider economic, environmental, and social impacts.

REFERENCES

Al-Khalidi, H., and A. Kalam. 2006. The Impact of Underground Cables on Power Transmission and Distribution Networks. First International Power and Energy Conference. PECon 2006. November 28-29, 2006, Putrajaya, Malaysia. 575-579 [online]. Available: http://ieeexplore.ieee.org/stamp/stamp.jsp?tp=&arnumber=4154561 (accessed December 19, 2012).

Bobrick, B. 1981. Labyrinths of Iron: A History of the World's Subways. New York: Newsweek Books.

Boyle, C., G. Mudd, J.R. Mihelcic, P. Anastas, T. Collins, P. Culligan, M. Edwards, J. Gabe, P. Gallagher, S. Handy, J.-J. Kao, S. Krumdieck, L.D. Lyles, I. Mason, R. McDowall, A. Pearce, C. Riedy, J. Russell, J.L. Schnoor, M. Trotz, R. Venables, J.B. Zimmerman, V. Fuchs, S. Miller, S. Page, and K. Reeder-Emery. 2010. Delivering sustainable infrastructure that supports the urban built environment. Environmental Science and Technology 44(13):4836-4840.

Bumby, S., E. Druzhinina, R. Feraldi, D. Werthmann, R. Geyer, and J. Sahl. 2010. Life cycle assessment of overhead and underground primary power distribution. Environmental Science and Technology. 44(14):5587-5593.

Canto-Perello, J., J. Curiel-Esparza, and V. Calvo. 2009. Analysing utility tunnels and highway networks coordination dilemma. Tunnelling and Underground Space Technology 24(2):185-189.

CGA (Common Ground Alliance). 2008. Best Practices Guide 5.0. Alexandria, VA: CGA.

CNRC (National Research Council Canada). 2005. Municipal Infrastructure Investment Planning Report: Social Cost Considerations for Municipal Infrastructure Management. Report B-512.8. Ottawa, Ontario: CNRC.

De Saventhem, E.M. 1977. Insuring risks underground - some general considerations. Underground Space. 2(1):19-26.

DOT (U.S. Department of Transportation). 2007. Statewide Transportation Planning; Metropolitan Transportation Planning; Final Rule, February 14. Fed. Reg. 72(30):7224-7286.

Elkington, J. 1994. Towards the sustainable corporation: Win-win-win business strategies for sustainable development. California Management Review. 36(2):90-101.

Elkington, J. 1997. Cannibals with Forks: The Triple Bottom Line of 21st Century Business. Oxford: Capstone.

EPA (U.S. Environmental Protection Agency). 2009. Endangerment and Cause or Contribute Findings for Greenhouse Gases Under Section 202(a) of the Clean Air Act; Final Rule. Federal Register 40 CFR Chapter 1. December 15 [online]. Available: http://www.epa.gov/climatechange/Downloads/endangerment/Federal_Register-EPA-HQ-OAR-2009-0171-Dec.15-09.pdf (accessed July 22, 2012).

EPA. 2010. National Environmental Protection Act [online]. Available: http://www.epa.gov/compliance/nepa/ (accessed August 3, 2010).

Ewing, R. 1997. Is Los Angeles-style sprawl desirable? Journal of the American Planning Association. 63(1):107-126.

Gilchrist, A., and E.N. Allouche. 2005. Quantification of social costs associated with construction projects: State-of-the-art review. Journal of Tunnelling and Underground Space Technology. 20(1):89-104.

Gordon, P., and H.W. Richardson. 1997. Are compact cities a desirable planning goal? Journal of the American Planning Association. 63(1):95-105.

Guinée, J.B. ed. 2002. Handbook on Life Cycle Assessment, Operational Guide to the ISO Standards. Dordrecht, The Netherlands: Kluwer.

Haimes, Y.Y. 2004. Risk Modeling, Assessment, and Management. Vol. 30. Hoboken, NJ: John Wiley and Sons.

Handy, S., X. Cao, and P.L. Mokhtarian. 2005. Correlation or causality between the built environment and travel behavior? Evidence from Northern California. Transportation Research Part D-Transport and Environment. 10(6):427-444.

Hendrickson, C.T., L.B. Lave, and H.S. Matthews. 2006. Environmental Life Cycle Assessment of Goods and Services: An Input-Output Approach. Washington, DC: Resources for the Future Press.

IFC Consulting. 2003. Overview of the Potential for Undergrounding the Electrical Networks in Europe. Report Prepared for DG/TREN (Director-General for Transport and Energy) European Community, by IFC Consulting, London, UK [online]. Available: http://ec.europa.eu/energy/gas_electricity/studies/doc/electricity/2003_02_underground_cables_icf.pdf (accessed June 12, 2012).

ISO (International Organization for Standardization). 1997. ISO 14040. Geneva, Switzerland: ISO.

ISO. 2006. ISO 14044. Geneva, Switzerland: ISO.

ITA WG13 (International Tunneling and Underground Space Association Working Group Number 13). 2004. Underground or aboveground? Making the choice for urban mass transit systems. Tunnelling and Underground Space Technology. 19:3-28.

Jeon, C.M., and A. Amekudzi. 2005. Addressing sustainability in transportation systems: Definitions, indicators, and metrics. Journal of Infrastructure Systems. 11(1):31-50.

Johnson, B.W. 2006. Out of Sight, Out of Mind? A Study on the Costs and Benefits of Undergrounding Overhead Power Lines. Report Prepared for Edison Electrical Institute, Washington, DC [online]. Available: http://www.woodpoles.org/documents/UndergroundReport.pdf (accessed June 12, 2012).

Kagan, H.A., D.J. Leary, and G.E.K. Pratter. 1986. Design engineers' responsibilities during construction. Journal of Construction Engineering and Management. 112(3):394-402.

Li, X.Q., B. Chavez, D. Ocenosak, and R. Williams 2009. Building a Large-diameter Deep Sewer Pipeline under Urban Constraints. Paper E-2-04. Proceedings of the International No-Dig Show Conference, March 29-April 3, 2009, Toronto, Canada. Liverpool, NY: North American Society for Trenchless Technology.

Matthews, H.S., C.T. Hendrickson, and C.L. Weber. 2008. The importance of carbon footprint estimation boundaries. Environmental Science & Technology. 42(16):5839-5842.

MADOT (Massachusetts Department of Transportation). 2012. The Big Dig. Massachusetts Department of Transportation-Highway Division [online]. Available: http://www.massdot.state.ma.us/highway/TheBigDig.aspx (accessed June 15, 2012).

MWRA (Massachusetts Water Resources Authority). 2006 . Metropolitan Boston's Water System History [online]. Available: http://www.mwra.state.ma.us/04water/html/hist1.htm (accessed January 3, 2011).

Najafi, M. 2005. Trenchless Technology: Pipeline and Utility Design Construction and Renewal. New York: McGraw-Hill.

NAE/NRC (National Academy of Engineering and National Research Council). 2003. Completing the "Big Dig": Managing the Final Stages of Boston's Central Artery/Tunney Project. Washington, DC: National Academies Press.

NRC. 2010. Hidden Cost of Energy: Unpriced Consequences of Energy Production and Use. Washington, DC: National Academies Press.

O'Rourke, T.D. 1983. Problems of size and complexity in urban underground projects. Underground Space 7(3):182-186.

O'Toole, R. 2009. The Myth of the Compact City: Why Compact Development Is Not the Way to Reduce Carbon Dioxide Emissions. Policy Analysis No. 653. Washington, DC: CATO Institute [online]. Available: http://www.cato.org/publications/policy-analysis/myth-compact-city-why-compact-development-is-not-way-reduce-carbon-dioxide-emissions (accessed June 13, 2012).

OMB (The White House Office of Management and Budget). 2010. Guidelines and Discount Rates for Benefit-Cost Analysis of Federal Programs. Circular A-94 [online]. Available: http://www.whitehouse.gov/sites/default/files/omb/assets/a94/a094.pdf (accessed August 2, 2010).

Palmer, T.C. 2004. For property owners, parks mean profits. The Boston Globe, June 14, 2004 [online]. Available: http://www.boston.com/news/local/massachusetts/articles/2004/06/14/for_property_owners_parks_mean_pro64257ts/ (accessed June 13, 2012).

Parker, H.W. 2007. Risk Analyses and Life Cycle Costs of Underground Facilities. Presentation at the ISRM-ITA Specialized Session, 11th Congress of the International Society for Rock Mechanics, July 9-13, 2007, Lisbon, Portugal.

Parsons Brinkerhoff. 2012. Electricity Transmission Costing Study: An Independent Report Endorsed by the Institution of Engineering & Technology (issued 31 January 2012; with errata April 2012).

Pucker, J., E.N. Allouche, and R.L. Sterling. 2006. Social Costs Associated with Trenchless Projects: Case Histories in North America and Europe. Paper C-4-04. Proceedings of the 2006 NASTT No-Dig Conference, March 26-31, Nashville, TN. Arlington, VA: North American Society for Trenchless Technology (NASTT).

Reijnders, L. 1996. Environmentally Improved Production Processes and Products: An Introduction. Berlin: Kluwer Academic Press.

Rinaldi, S. M., J. P. Peerenboom, and T. K. Kelly. 2001. Identifying, understanding, and analyzing critical infrastructure interdependencies. IEEE Control Systems 21(6):11-25.

Schwanen, T., M. Dijst, and F.M. Dieleman. 2004. Policies for urban form and their impact on travel: The Netherlands experience. Urban Studies. 41(3):579-603.

Seo, J.W., and H.H. Choi. 2008. Risk-based safety impact assessment methodology for underground construction projects in Korea. Journal of Construction Engineering Management. 134(1):72-81.

Shammin, M.R., R.A. Herendeen, M.J. Hanson, and E.J.H. Wilson. 2010. A multivariate analysis of the energy intensity of sprawl versus compact living in the U.S. for 2003. Ecological Economics. 69(12):2363-2373.

Sihabuddin, S., and S.T. Ariaratnam. 2009. Methodology for estimating emissions in underground utility construction operations. Journal of Engineering, Design, and Technology. 7(1):37-64.

Sterling, R.L. 2005. Urban underground space use planning: A growing dilemma. In Underground Space and Rock Mechanics: Proceedings of the 10th ACUUS International Conference-Underground Space: Economy and Environment, January 24-28, 2005, Moscow.

Sterling, R., H. Admiraal, N. Bobylev, H. Parker, J.P. Godard, I. Vähäaho, C.D.F. Rogers, X. Shi, and T. Hanamura. 2012. Sustainability issues for underground spaces in urban areas. Proceedings of ICE - Urban Design and Planning [online]. Available: http://www.icevirtuallibrary.com/content/article/10.1680/udap.10.00020.

Taylor, A.W. 2008. Alaska Way Viaduct: Evaluation of Seismic Retrofit Options. Final Report of the Alaskan Way Viaduct Independent Project Management Team. Seattle, WA: KPFF Consulting Engineers [online]. Available: http://www.wsdot.wa.gov/NR/rdonlyres/C4C966C1-2723-4187-AD4D-0F3FA1875C19/0/KPFFretrofit080925.pdf (accessed June 15, 2012).

TRB (Transportation Research Board). 2009. Driving and the Built Environment: The Effects of Compact Development on Motorized Travel, Energy Use, and CO2 Emissions – Special Report 298. Washington, DC: National Academies Press.

UN (United Nations). 1999. Technical Report on Space Debris. Text of the Report adopted by the Scientific and Technical Subcommittee of the United Nations Committee on the Peaceful Uses of Outer Space. New York: UN [online]. Available: http://orbitaldebris.jsc.nasa.gov/library/UN_Report_on_Space_Debris99.pdf (accessed June 15, 2012).

Weber, P., G. Medina-Oliva, C. Simon, and B. Iung. 2012. Overview on Bayesian networks applications for dependability, risk analysis, and maintenance areas. Engineering Applications of Artificial Intelligence 25(4):671-682.

Winqvist, T., and K.E. Mellgren. 1988. Going Underground. Stockholm: Royal Swedish Academy of Engineering Sciences. 191 pp.

Woodroffe, N.J.A., and S.T. Ariaratnam. 2008. Cost and Risk Evaluation for Horizontal Directional Drilling versus Open Cut in an Urban Environment. Practice Periodical on Structural Design and Construction. 13(2):85-92.

6

Innovative Underground Technology and Engineering for Sustainable Development

Geotechnologies and related science and engineering fields make it possible to use underground space to support livable, resilient, and sustainable cities. Geotechnical applications have supported the design and construction of underground facilities, and will continue to be critical to the delivery of underground facilities with lower initial costs and risk, and better lifecycle performance. To contribute to a more resilient and sustainable future, geotechnology will need to more closely integrate the many disciplines related to site investigation, design, construction, operation, and risk management of underground facilities. A better understanding of the sustainability of underground use—for example, minimizing deterioration, increasing resilience, making holistic decisions concerning subsurface hydrogeologic and thermal environments—also will be necessary. Improvements in underground technologies have enabled great strides in urban development in recent decades, but the complexity and unpredictability still inherent in underground construction are indications that many challenges remain.

This chapter provides a brief overview of the state of the art in various technologies that support underground construction and facility operation. Highlighted are technologies that provide opportunities for significant improvement in the delivery of cost-effective lifecycle performance for underground facilities, contribute to improvements in underground space usage, and contribute to resilient and sustainable urban solutions.

EVOLUTION OF TECHNOLOGY

Technological innovation can advance engineering practice and increase the appeal of underground space. Technological and engineering advances have always been crucial to efficient and economical underground development. Many technological developments have been motivated by practical challenges encountered during construction of a project (e.g., the development of the tunnel shield by Brunel), and the tunneling industry has contributed to or been instrumental in many of these. The highly automated modern tunneling boring machine is an example of an an industry led development as are water proofing and ground improvement technologies that have been introduced and popularized. In close partnership with academia, industry has developed many analysis and design tools (e.g., finite element analysis methods).

Since the time of the Pharaohs, tunnels have been built by cut-and-cover construction methods (El Salam, 2002). The invention of the tunnel shield—which supports unlined ground to reduce the risk of collapse, Sir Marc Isambard Brunel and his son Isambard Kingdom Brunel were able to excavate a tunnel under the Thames River (London) between 1825 and 1843 (Muir Wood et al., 1994; Skempton and Chrimes, 1994) (see Figure 6.1 for a drawing of Brunel's shield). Previous projects involving tunnel boring in soft, saturated soils had been extremely difficult or impossible to complete. The tunnel created an important connection between the north and south banks of the Thames that is still in use almost 170 years later. The application of this new technology heralded the era of shield tunneling.

Electrically powered locomotives ushered in the era of modern subway systems around the turn of the twentieth century. Electrification alleviated concerns about hazardous diesel or coal fumes and allowed long-distance underground train travel. Innovations in large-scale ventilation systems permitted underground roadway development. Climate control systems, improved lighting, and more effective signage made the underground environment more hospitable, comfortable, and appealing for retail functions and mass transportation. Advances in materials technology, computer science, robotic construction technology, and laser guidance have allowed improved subsurface excavation using modern slurry shield and earth pressure balance boring machines[1] (Figure 6.2) and rock tunnel boring machines (Figure 6.3). Those technologies made it feasible to construct tunnels exceeding 50 kilometers in length and at diameters approaching 20 meters, and to tunnel under challenging geologic conditions (e.g., in soft flowing ground or highly fractured rock under high ground and water pressures). Ground modification technologies—e.g., injecting cementitious agents to strengthen and reduce permeability of soil and rock, or temporarily freezing of water-bearing

[1] Slurry shield and earth pressure balance boring machines for boring in saturated soils are designed to withstand water under pressure.

FIGURE 6.1 Brunel's tunnel shield. Marc Isambard Brunel developed this tunnel shield technology to build the first subaqueous tunnel beneath the Thames River (1821-1825). Brick walls were built at the faces of the tunnel and held in place while alternate shields were pushed forward 6 inches. The completed tunnel was 38 feet wide and accommodated two carriageways. SOURCE: http://en.wikipedia.org/wiki/File:Thames_tunnel_shield.png (accessed June 27, 2012). Public Domain.

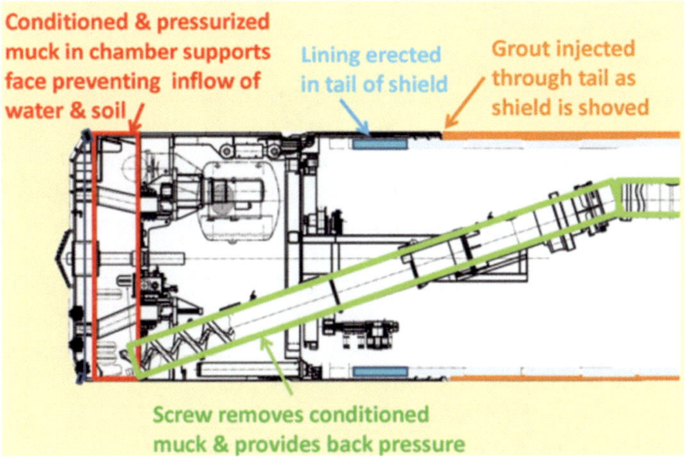

FIGURE 6.2 Cross-section of earth pressure balance tunnel boring machine. This tunneling technology is ideal for homogenous soft soils. A screw conveyor is used to transport spoil from the face and helps to control pressure with the coordinated advancement of the machine. The excavation chamber is filled to support the face and allow the machine to be reactive to earth and groundwater pressures. SOURCE: E.J. Cording.

FIGURE 6.3 Cutter head of a rock tunnel boring machine used to excavate the Chattahoochie tunnel. Disk cutters cut grooves approximately 4 inches apart in this example. SOURCE: E.J. Cording.

materials (Figure 6.4)—broadened the geologic and hydrologic conditions under which underground construction may occur. Horizontal directional drilling revolutionized installation of many utilities and greatly reduced the need to close streets to traffic and disrupt life in urban situations.

Many of the technologies described above led to changes in engineering practice, and in some cases, to new paradigms in urban planning. Similarly, today's engineering and technology developments will be crucial to an economically constructed, functional, attractive, energy efficient, and sustainable urban environment. This chapter is grouped under the following themes:

- technologies for underground site characterization, including geologic setting, rock and soil properties, and existing underground infrastructure;
- technologies for design and analysis for underground technologies;
- technologies for construction of underground space;
- technologies for effective asset management; and
- technologies that promote sustainability and resilience.

These themes are not necessarily sequential or independent. Observations made during the application of each may inform decision making during any phase of development or operation. Infrastructure design may identify further site characterization needs, and unanticipated conditions encountered during construction

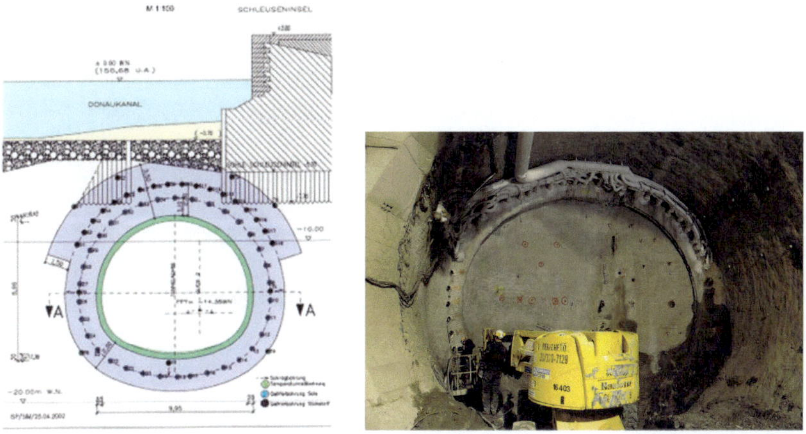

FIGURE 6.4 Artificial ground freezing is applied in the excavations extending under the Danube River. Freezing groundwater around an excavation improves the load carrying capacity of the soils and provides temporary support during construction. (Left) A cross-sectional diagram with the locations of the freezing pipes indicated. SOURCE: IMWS, 2009. Reprinted with permission. (Right) Photograph showing excavation with freezing pipes in place. SOURCE: http://www.tunneltalk.com/images/BudapestMetro/6-Budapest-Metro-GroundFreezingApplied.jpg.

may require revisions in design. Monitoring and characterization ideally should occur throughout the infrastructure life cycle. Observations can lead to a type of informed decision making called observational method (e.g., Peck, 1969; Institute for Civil Engineers (Great Britain), 1996) that can improve economy and safety. Many geotechnical engineers refer to the framework for this method originated by Peck and described in a publication by Nicholson and others (1999).

The discussions within each theme are illustrative of technologies in use or that have significant potential for the future. By their nature, disruptive technologies are difficult to anticipate, but can fundamentally shift the way underground space is developed and used. Many of the technologies described in this chapter depend on the use of the observational method for effective decision making. Suggested are technologies for analyses that allow improved application of the observational method.

TECHNOLOGIES FOR UNDERGROUND SITE CHARACTERIZATION

Engineering urban underground space requires detailed knowledge of the underlying geology and the geologic and human-development histories of a site, alignment, and adjacent areas that may affect or be affected by proposed development. Better subsurface characterization supports better decision making. Minimizing unanticipated ground conditions may allow optimized design and more judicious use of resources during construction. Detailed understanding of how the site relates to the broader natural and urban systems allows more complete understanding of the existing engineering challenges and informs underground infrastructure locations and alignments, design, choice of construction methods and tools, and long-term operation of the facility and adjacent structures.

Site characterization activities often begin through study of existing data and published information. Currently, relevant information must be gathered from many sources and may not provide adequate—or accurate—information about the geological setting or existing underground structures. Ongoing advances in computational capabilities (e.g., massive database systems, data mining) and georeferencing of data (e.g., survey-grade global positioning systems and geographical information systems) could be of great use in the future. These issues are further explored later in this chapter.

Characterizing Geology, Geologic Material Properties, Contamination, and Natural Hazards

A geologic perspective in site characterization is necessary to appreciate, quantify, and manage uncertainties in and variability of soil and rock properties and behavior (e.g., composition, stress-strain behavior, permeability, abrasivity, thermal conductivity) and groundwater conditions, ultimately minimizing costs or avoiding overly conservative design (NRC, 2006; FHWA, 2009). Preliminary field and laboratory work supports preliminary project design and construction planning, but further characterization narrows uncertainties and provides detail on important geologic features (e.g., boundaries of geologic units, fault zones), supports design of specific underground works (e.g., shafts, rooms), and provides information related to other special requirements (e.g., avoiding environmental contamination or in situ stress evaluation). The natural underground environment is inhomogeneous, anisotropic, and highly variable over short spatial extents. Large volumes of geologic material often must be characterized in great detail to detect stratigraphic changes and discontinuities important for predicting ground response to construction. A broad range of invasive and noninvasive technologies and tools are available to carry out in situ field investigations (see for example, FHWA, 2009), but existing assessment tools cannot provide complete spatial coverage, accurate zonation, and in-situ material properties.

At a project scale, hazardous materials encountered during underground construction can add large and unexpected costs to a project and delay project delivery. Characterizing natural and anthropogenic hazardous materials (e.g., chemical contamination and radiation) and their effects on the natural and built environments for particular construction and operation activities is vital. Understanding any hazardous materials that may be released or transported as a result of construction and operation is important to long-term sustainability and resilience. Similarly, identifying adequate design approaches to protect underground infrastructure from natural hazards such as earthquakes or floods is critical to a resilient and well-functioning underground facility. Because these topics have a broad connection to sustainability and resilience, their characterization is discussed later in the chapter.

Environmental concerns that extend to underground storage and disposal also need to be considered. For example, society is grappling with the risks associated with the emerging technology of carbon dioxide sequestration. As more large-scale sequestrations are planned, the need to examine their potential impacts on the ability to develop underground space becomes even greater, because, for example, carbon dioxide could seep into underground space. Therefore, the solution of one problem could inadvertently result in another problem. A recent NRC report explored the risk associated with induced seismicity as a result of carbon capture and storage and makes specific research recommendations related to, for example, factors other than pore pressure that influence seismicity, and development of physiochemical and fluid mechanical models for carbon dioxide injection into potential underground storage reservoirs (NRC, 2012).

Choice of characterization tools depends on a number of factors including depth of interest and ground conditions (e.g., soil versus rock; saturated versus partially saturated). Both traditional in situ technologies (e.g., direct measurement) and noninvasive technologies (e.g., geophysical) can be used to characterize natural and manmade features. Some construction sectors provide guidance on site characterization technology choices through extensive lists of tools and techniques (e.g., FHWA, 2009). Training and experience in the proper use of the tools, however, is usually as critical as the choice of technology itself.

Invasive Technologies

In situ testing tools provide direct physical measurement of material properties. In soils, for example, standard penetration tests and cone penetrometer tests (e.g., electric, piezocone, and seismic tools) are used to sample or test soil layers directly by drilling or thrusting sampling tools into the ground. Rock sampling and testing can be borehole based or conducted on the removed core. Borings are used to characterize properties such as soil strength, stiffness, dynamic shear wave velocity, and groundwater properties and quality, and the geology at the borehole location. An individual boring may or may not represent the subsurface only a short distance away given the potential variation in geology. Directionally inclined and horizontal boreholes and oriented probing tools also can be used to investigate specific features or the distribution of materials. Horizontal probing allows exploration of the subsurface along the length of the alignment of a tunnel or other infrastructure. The use of oriented exploration tools, while common in the energy exploration industry, is less common for civil underground structure development. This may be due to cost, but perhaps also to unfamiliarity with the technique among site investigation professionals. It may be argued that incentives for efficiency are less evident for engineers than for contractors and owners who realize savings from efficiency.

Recent developments in boring technologies include cryogenic drilling capabilities for boring in difficult materials and measurement while drilling (MWD) systems that provide early information on the materials useful for guiding future borings and planning an efficient testing program. The profession has yet to widely adopt these techniques.

Noninvasive Technologies

Noninvasive site characterization tools include remote sensing (e.g., satellite and terrestrial light detection and ranging (LIDAR), digital photogrammetry, radiometric technologies, and interferometry methods) and ground geophysical techniques (e.g., seismic refraction and reflection, spectral analysis of surface waves (SASW), crosshole tomography, geoelectric, electromagnetic, and potential field methods [gravity, magnetic]) that provide data from which subsurface conditions may be inferred. Noninvasive techniques are best used in combination with invasive techniques to provide a more complete understanding of underground conditions. Advantages of noninvasive technologies are the speed with which they can be used and that larger volumes of the subsurface can be characterized. Disadvantages are that data generally must be reduced from their raw form—inversion modeling is often required to evaluate ground zonation and materials properties. Such models are non-unique (e.g., a single data set can yield infinite models), and hence special skills and knowledge are required to reduce and interpret data. The cost of some of these methods can also be high, but as the methods become more common and the technologies continue to improve (e.g., laser scanning), the cost of data acquisition and analysis will go down.

There is significant opportunity to improve data gathering related to ground properties and the presence and location of existing structures using noninvasive technologies, but there are physical limitations in terms of the scale of objects to be characterized and the material property differences that can be identified relative to the depth of investigation possible. There is, for example, a practical limit for pipe detection using surface-based ground penetrating radar (GPR), reported to be the ratio of approximately 12:1 for the detection depth to the detectable pipe diameter, even under favorable soil conditions (Sterling et al., 2009). This means that a 1-foot diameter pipe can only be detected if within 12 feet of the surface, and a 1-inch pipe can only be detected up to a depth of 1 foot. Research into the fusion of multi-sensor data that would allow noninvasive technologies to accommodate a wider range of ground conditions and to improve their ability to resolve ground properties and the presence and location of buried objects is under way both in the United States and overseas. Similarly, research by the military into the detection of land mines and deep covert tunnels can have significant benefits in broader civil engineering applications.

Characterizing Existing Infrastructure and Legacy Construction Materials

Failing to locate existing infrastructure before repairing existing or installing new infrastructure is a potential source of accidents. Legacy construction materials, including unmapped abandoned piles, foundations, or tiebacks that once provided support during previous construction, are regularly encountered during underground construction. Identifying and characterizing these artifacts is a necessary part of site characterization. Historical records found in planning departments can be used to identify locations of some legacy materials, but the records are often incomplete, inaccurate, or missing, thereby necessitating a greater reliance on exploration technologies, especially noninvasive, for characterization.

Unmapped or inaccurately mapped underground infrastructure poses potential hazards and risks for underground construction workers, the construction site, other infrastructure, and other people in the vicinity. Encountering unexpected infrastructure may necessitate revised construction planning or repairs. Existing or legacy infrastructure may be avoided, but sometimes must be protected with support designed to avoid displacement or damage to existing or planned infrastructure. The committee notes that every open-cut excavation, bore, or tunnel is an opportunity to assess and document the ground properties and structures encountered for present and future applications. The hundreds of thousands of open-cut excavations for utility work made every day in the United States, for example, offer repeated opportunities to collect and archive such data. However, the fidelity of nonivinasive techniques to identify subsurface infrastructure needs to be improved. Additionally, investigation technologies need to be integrated with new physical tools and administrative structures to capture this type of information. Mechanisms that allow dynamic archiving (e.g., continuous updating and modification) of these data are critical to the sustainability of urban infrastructure.

Interpreting and Integrating Site Characterization Data

Site characterization information and data must be processed and evaluated to develop interpretative geologic models and to generate the engineering parameters to be used in underground facility design (see examples in Box 6.1). Many field techniques used for preliminary property classification have been applied for decades and are subject to gross differences in interpretation. Typically, field classification cannot substitute for laboratory verifications. Many tools aid interpretation of, for example, rock classification including empirically based procedures such as the Q-system (for rock quality classification) (Barton et al., 1974), the Rock Mass Rating system (Bieniawski, 1976, 1989), and the Geological Strength Index (Hoek, 1994) (see Figure 6.5 for example classification). Classification schemes for characteristics such as strength and stiffness also are used to establish the input for advanced numerical analysis procedures.

BOX 6.1

Three-dimensional Geologic Modeling

Three-dimensional modeling of ground conditions that incorporate geotechnical and geophysical data is conducted extensively in the United States for development activities including resource exploration and extraction and infrastructure development. Similar modeling techniques can be applied to urban infrastructure planning, risk modeling, and resource management as is done by the British Geological Survey (see Figure 1). Such models can provide multiple "views" (e.g., orientations), be "exploded" (e.g., layers can be visually separated to isolate specific features or units), and otherwise manipulated to identify predicted physical properties at depth, the locations of anthropogenic structures, aquifer vulnerabilities, and other qualities. Such comprehensive

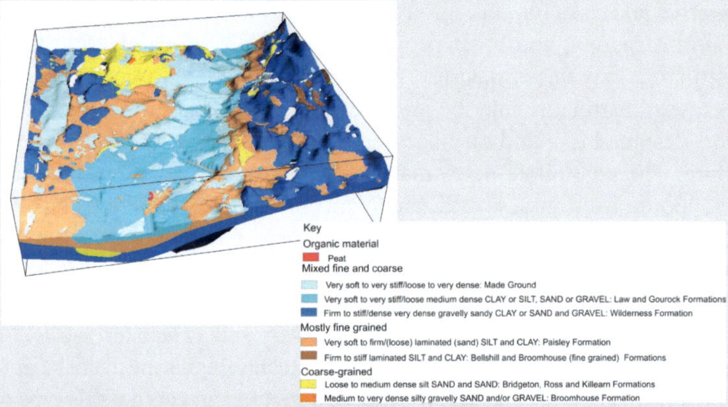

FIGURE 1 Example of three-dimensional engineering geological modeling employed by the British Geological Survey for visualizing variability in geologic materials and their physical properties. SOURCE: Reeves, 2010. Reproduced by permission of the British Geological Survey. © NERC. All rights reserved. CP12/073.

From a broad perspective, however, insufficient use is made of all the classification and material testing that is carried out on the thousands of individual projects that occur in a medium- to large-sized city every year. Collection and integration of such data remain difficult because, to be useful, the data must be carefully documented and referenced as to location, depth, other properties, and pedigree (e.g., data sources, what tests were run, and was test equipment properly calibrated). Also, and perhaps the most telling, are the significant disincentives for project owners and their consultants to release data because of concerns related to liability and loss of proprietary knowledge. Nevertheless, regulations exist, for example, that require well boring operations to submit their boring logs to state geological surveys. Steps to usefully capture more of the geotechnical views of data can enhance decision making and help to quantify uncertainties that historically are a source of difficulties in contracting and litigation. Confidence maps can be created based on data density and geologic complexity that indicate areas of low or high uncertainty in models (see Figure 2). Given variability of geologic conditions and the spatial limitations of underground characterization tools, information about the underground is often limited and includes significant uncertainties.

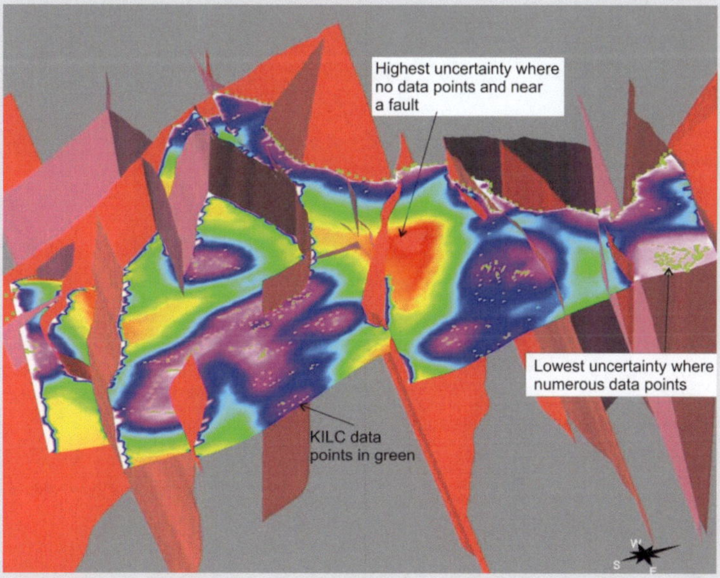

FIGURE 2 A bedrock confidence map produced for an area in Glasgow, Scotland. The green points represent actual data points; the roughly vertical planes represent faults. Contours represent levels of uncertainty based on data density and geological complexity of modeled surface (red indicating high uncertainty). Such maps provide valuable insights regarding where more data may be needed. SOURCE: Reeves, 2010. Reproduced by permission of the British Geological Survey. © NERC. All rights reserved. CP12/073.

data generated represent important ways to help enable the sustainability of the urban underground and the regions it serves.

To move toward engineering practices that are consistent with such sustainability goals, data related to underground infrastructure development need to be archived in formats and with tools that make them retrievable and accessible for the infrastructure life cycle—and beyond (to account for infrastructure artifacts in place well after closure or decommissioning). These issues will be discussed in a later section related to the critical challenges of archiving infrastructure-related data.

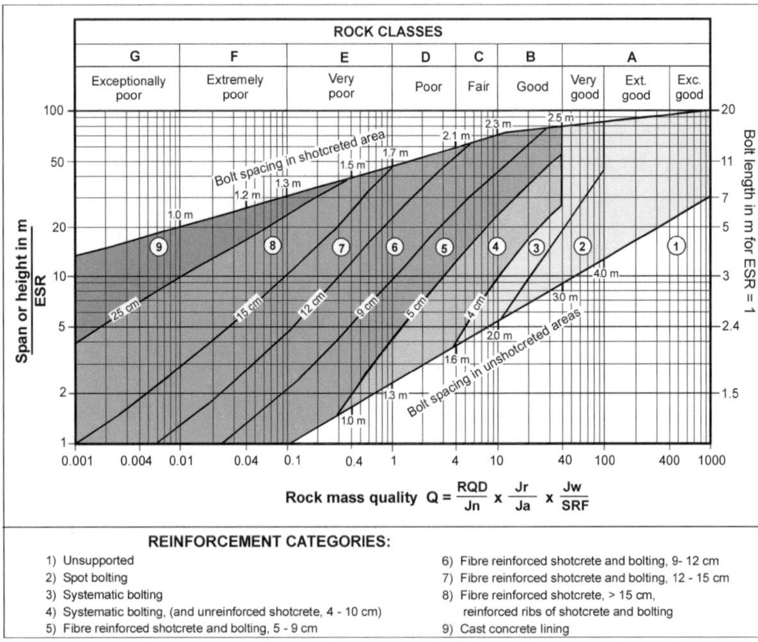

Figure 6.5 Support categories estimated from the tunneling quality index (after Grimstad and Barton, 1993) SOURCE: Palmstrom and Broch, 2006.

TECHNOLOGIES FOR DESIGN

Sustainability as an integral part of design is a relatively new concept. Sociopolitical and economic issues, discussed in earlier chapters, influence infrastructure design, perhaps at the expense of sustainability. Design and analysis of underground infrastructure is often heavily focused on the immediate opening and support of underground space; long-term issues related to sustainable maintenance and use are often overlooked, as are lifecycle contributions of the infrastructure to society. Project constraints such as rights of way and access can impact selection of project alignment, and physical constraints such as those associated with water conveyance gravity systems (e.g., slope, number of lift stations, length of drives), maximum grade (for construction and operation), and shaft location also may affect design choices. New technologies or policies that reduce the adverse affects of such physical constraints could reduce the cost of developing underground space.[2] As discussed in Chapter 4, codes and standards guide design and operation, especially where public safety is affected. Although building codes can protect the health and safety of those constructing, operating, or using infrastructure, building codes are static in nature and, therefore, can detract from sustainability goals and leave little room for flexibility or evolution of technologies.

Constraints created by the limitation of design tools, or by the limitations of those who use the tools, also limit the ability to design underground infrastructure optimally and to reduce lifecycle costs. Underground infrastructure design can benefit from iterative analyses of designs—especially innovative designs—in a virtual environment. Design performance in difficult ground conditions can be predicted using numerical models, and results of analyses can inform infrastructure design modifications that meet constraints and desired parameters. The process can be repeated as often as necessary until desired model responses and optimal design are achieved. Commercial software packages are available for this sort of analysis, but their use is successful only if their limitations are recognized. Needed in geotechnical design is greater consistency in numerical modeling usage, project design validation, and integration of modeling with other analysis components (scaling) to arrive at more sustainable design. This requires software enhancement, especially with respect to graphical user interface interactions, error checking, and user training. Another challenge is the lack of appropriate ground behavior model input. Field and laboratory test data are often not appropriate for input in advanced analysis or model verification. Stronger and more direct linkages between field investigation data and model development are needed (e.g., Hashash et al., 2009).

The proper combination of technology, training, and skills are vital to good engineering design. The design of sustainable underground facilities is a complex and iterative process that is necessarily a team effort, as described in Chapter 3. Continuous communication between engineers, owners, and interested third parties is necessary to develop optimal designs that meet the owner's specific requirements and societal sustainability goals. Input from a broader array of stakeholders helps create infrastructure that is integrated into the system of systems that comprise urban development. Designers need to contend with how the space fits within existing infrastructure, its long-term impacts on the total urban environment, and its resilience over time.

The next sections describe typical major inputs to the design of underground projects.

[2]An example of this is a Japanese law pertaining to the public right to space of greater than 40 m depth as a means to enhance planning and sustainability of underground resources. The laws allows rational choices for project alignments and preservation of the underground for future uses (Konda, 2003).

Integrating Design and Analysis

Good engineering design includes analysis of system performance as part of design development. Many tools are available to design details of underground structures—for example, thickness of supports or appropriate size and location of ventilation shafts. Empirical and analytical design methods, ideas and procedures based on experience, building codes, and analytical and computational software packages all contribute to optimal design of underground structures. However, their effectiveness directly correlates with the quality of data input and the knowledge base and experience of those applying the tools.

Experience

The lack of education and training that allows good engineering of underground systems is of growing concern in the United States and will ultimately affect the resilience of underground infrastructure and sustainability of the urban development. Fewer than five U.S. institutions of higher education provide student opportunities in underground construction, and engineering talent is being imported from overseas. Similarly, it is critical for the underground construction industry to develop mechanisms that allow the relevant transfer of experience from senior personnel to young engineers.

Empirical Design Tools

Empirical design methods rely on quasi-quantitative approaches to material characterization based largely on experience-driven judgments. They are not necessarily based on fundamental mechanisms of ground and structure behavior. For example, soil and rock classification systems are based on the comparison of observed material behaviors from the infrastructure site to similar observations made elsewhere. Behavior is predicted based on those comparisons, but the underlying reasons for the behaviors may not be well understood. Even so, the soil and rock classifications based on empirical characterization often inform support requirements and design specifications. In some cases, empirical characterization may be based on a large number of prior observations, and resulting conclusions may be robust. In other cases, few data exist to enable comparisons, and empirical characterizations may be more similar to "best guesses." Empirically based conclusions can be considered first-order estimations, and further observations may be necessary to confirm or refine those estimates. In other words, they can be used to interpolate but not extrapolate. Improper use of empirical tools can result in safety hazards, poor performance, or unnecessary expenses. Educating professionals about the limitations of empirical methods is one way to improve their use. Improving the databases of observations and the methods for their expansion is another.

Analytical and Computational Methods

Analytical methods—primarily closed-form mathematical solutions that do not require the use of a computer—for calculating stresses, strains, ground movements, groundwater flow, and other properties are critical to improving the understanding of underground structure behavior in concert with the ground environment. They provide data on the responses of underground structures for particular idealized boundary conditions and material behaviors. Analytical methods are key tools for building knowledge and judgment regarding geotechnical behavior, for carrying out preliminary design studies (or full design given suitable configurations), and for comparing complex computational model outputs to gauge reliability. However, they cannot capture the full complexity of most underground structures and geological conditions.

Computational methods such as finite element, finite difference, and discrete element methods are used to estimate behaviors of soil, rock, or engineered structures when a sufficient amount of detailed information is available—for example, to refine design based on responses—but can be useful when there is limited prior experience working under specific conditions. Behaviors may be predicted using computer software programs. Computational methods may be used to estimate, for example, structure-ground interactions and deformations, changes in the ground environment (e.g., thermal changes and groundwater flow and contamination), and the propagation of fire and smoke in occupied underground spaces. All computational methods, whether based on discretization (e.g., differential versus integral) or continuity assumptions (e.g., continuum, discrete rock blocks, or some hybrid), require knowledge of spatial information and materials properties (e.g., intact materials, discontinuities, and fluid properties).

In many respects, the rapid development of computational methods during the past several decades has moved beyond present ability to gather sufficiently detailed data to populate the models or validate complex model output (e.g., behaviors) before model information is incorporated into design. The ability of a designer to compare the results of different analytical and simulation approaches and to access relevant case history data could be significantly enhanced. To benefit more fully from advances in computational methods, ground information models can be linked dynamically with data from excavation equipment, support systems, and updated simulations to provide useful feedback loops. The information can be used during construction to validate predictive models used in design, improve ground characterization, and update expected ground and structure responses. Similarly, data collection from the ground environment and the underground infrastructure itself throughout the infrastructure life cycle would enhance understanding of the sustainability implications of the structure (e.g., impacts to and from the surrounding environment). The United States has been the leader in the development of numerical simulation software for geotechnical engineering for several decades, but that lead is threatened by advances in

BOX 6.2

Analytical and Computational Model Development

More and better data are not enough. Better models are needed to investigate the behavior of complex adaptive and coupled infrastructure systems including

- analytical and computational models that predict and learn system performance, identify vulnerabilities, and provide a platform to investigate opportunities to increase security, robustness, resilience, capacity and efficiency, and cost-effectiveness;
- linkage elements and algorithms that allow simulation and exploration of cross-sector interactions and interdependencies; and
- tools to visualize and communicate the outputs of these advanced simulations.

Enhanced analytical techniques need to be applicable under routine operating conditions, and also need to capture emergent operating conditions and behaviors resulting from complex system interactions and non-routine internal or external stresses (e.g., earthquakes, major storms, sea level changes, terrorist activities).

Validated models could provide a platform to understand how best to

- integrate and operate above- and belowground infrastructure facilities for reliable and sustainable service provision;
- capture the potential of new technologies and materials to enhance performance of existing systems and to gain insight into the changing balances between centralized and decentralized system attributes;
- design and deploy sensing networks to efficiently and economically capture complex system behavior and improve system operation and security;
- develop long-duration research test beds for deployment of sensors in real environments;
- enable reliable data security for transmitted and stored data and develop resilient system architecture that can detect manipulated data; and
- explore new concepts in system design, operation, and maintenance that decrease vulnerability and provide flexibility and resilience.

simulation approaches in other countries. Multiple software packages now in use in geotechnical practice have been developed outside the United States. Box 6.2 describes some of the needs related to analytical and computational model development.

TECHNOLOGIES FOR UNDERGROUND CONSTRUCTION

Technological developments related to excavation, ground modification, improvement, support, tunnel boring, and use of excavated materials are important for developing underground space more efficiently, cost effectively, and under increasingly difficult ground conditions. Underground technology development and engineering advances are spurred by underground construction requirements throughout the world (Parriaux et al., 2006), and by the need to remain competitive in an international market for underground construction contracts. The importance of underground infrastructure development that contributes to sustainable development is now being recognized (e.g., ECTP, 2005).

Sustainability of the urban environment is partially dependent on minimizing disruptions to economic output of populated areas during construction (and later maintenance) of underground infrastructure. Underground projects often take years to complete, and during that time equipment is moved to and from an excavation site and huge volumes of excavated soils and rock (muck) must be moved. Streets may be blocked for extended periods, and truck movement on surface streets can be disruptive and polluting. Because each project and environment is unique, no single engineering solution minimizes disruptions, use of nonrenewable resources, and costs for all underground projects. However, technologies are available if there is a willingness to incur additional startup expenses. Advances in technologies such as new tunnel boring machines (TBMs) and trenchless technologies can minimize disruption, for example by allowing excavation in difficult ground conditions without open trenching.

The next sections provide examples of technologies used in the construction of underground space, including cut-and-cover tunneling and excavation, tunneling and ground support in rock and soils, ground improvement, and monitoring.

Cut-and-Cover Tunneling and Excavation

Open-cut excavation is a common and well-proven technique for constructing shallow (e.g., less than 150 feet [50 meters] deep) tunnels, earth-covered buildings, and basements and building foundations. Utility excavation also commonly uses open-cut trenching. Even bored tunnels or immersed tube tunnel sections will use cut-and-cover tunneling for access, support structures, and launch of boring machines. Open-cut technologies may be economically optimal for areas with few constraints on surface use, but can be quite disruptive in populated areas.

Open-cut excavation can accommodate changes in tunnel width and nonuniform shape and is often adopted for the construction of chambers and stations. Selection among available construction techniques such as cast-in place, bottom-up, or top-down, depends on ground conditions, available space, and environmental requirements (for example, Bickel et al., 1996). Excavations are normally braced and open to the surface with subsequent backfill or continued vertical structure construction (see Figure 6.6). In public rights-of-way and when using top-down methods, the excavation may be covered at an early stage of construction to allow traffic flow to resume or to allow construction of the upper

FIGURE 6.6 Braced excavation for a cut and cover tunnel section of the Boston Central Artery. SOURCE: Y. Hashash.

floors of a building while the basement excavation is continuing. Cut and cover also can be used in submarine tunnels.

Major cut-and-cover projects in urban areas provide opportunities to update other underground facilities in their vicinity. For example, utilidors may be incorporated in transportation projects, and underground pedestrian networks in city center redevelopments (as occurred in both Montreal and Toronto) may be built. The direct and indirect comparative impacts of cut-and-cover construction in the public right-of-way as compared to less intrusive construction approaches such as bored tunnels and trenchless technologies are not always well enough understood to inform decision making.

Tunneling and Ground Support Technologies in Rock

Rock excavation is required for all types of underground uses including mountain tunnels, mining operations, and large underground caverns where rock is at or near the surface. In historic times, rock excavation was painfully slow, relying on hand tools or even thermal shock[3] to disintegrate the rock. The introduction of blasting technologies in the sixteenth and seventeenth centuries revolutionized the speed of rock excavation and increased the range of projects that could be undertaken cost-effectively. In the past half-century, hard-rock TBMs have found widespread application creating medium length to long tunnels. Rock

[3]Fires were used to heat rocks, which were then quickly cooled with water. The rapid change in temperature caused rocks to crack and break. Broken rock could then be excavated.

excavation also benefited from the emergence of rock mechanics as a distinct branch of geotechnical engineering in the 1950s. Study of rock fragmentation, understanding of the post-peak strength behavior of rocks, and characterization of the influence of rock discontinuities on rock mass behavior are among the important outcomes from research in rock mechanics.

Drill and Blast Technologies

In drill and blast excavation, explosives and timed detonators are placed into drilled holes, the blast is carried out, the opening is ventilated, waste spoil is transported away from the excavation face, and the process starts again until the desired opening is obtained (e.g., Sellers et al., 2010). Blasting is typically the preferred solution from a cost perspective for short rock tunnels and for the excavation of foundations and rock caverns. Blasting, however, can be especially challenging in urban environments given higher amplitude vibrations (and therefore, ground shaking), larger sized spoil material, and a greater need for primary tunnel support than, for example, that result from TBM use. Excavation speeds are relatively slow, the process is noisy, and overbreak (excavation of too much material) is possible.

Alternatives to conventional drill and blast technologies are possible. For example, propagating fractures from boreholes in rock using controlled gas expansion could allow continuous excavation without releasing the excess energy that causes vibration and flying rock. The commercial application of such technologies so far has been restricted to specialized applications where conventional blasting is not feasible.

Blast design (e.g., blast patterns and sequences) could be more effective with improved computational models of drilling, blasting, vibrations, and displacements. Designs that minimize noise and vibration, for example, may allow blasting at any time of day, increasing productivity, minimizing disruptions to local neighborhoods, and minimizing damage to surrounding infrastructure. As cities become more densely developed and use more underground space, blasting approaches that are proven to minimize damage and nuisance will be desired.

Tunnel Boring Machine Technologies for Rock Excavation

Early variations of TBMs for soft rock were developed in the late 1800s (e.g., the Beaumont English boring machine was used on one of the early attempts to create the Channel Tunnel between England and France; Maidl et al., 2008). Early advances in TBM design were made by the Robbins Company of Seattle, Washington, which remained the world-wide leader in TBM design and manufacture for many decades. During the past decade, however, TBM manufacture has become increasingly dominated by non-U.S. companies.

Machine tunneling in harder rock remained a challenge because of inefficiencies and rapid wear of the pick-type rock cutters in use at the time. With the advent of full-face machine design and disc cutters for rock and proof of the value of the TBM approach (e.g., the Chicago Tunnel and Reservoir Plan), the hard-rock TBM using disk cutters gradually supplanted drill and blast technologies for longer tunnels.

Advance rates using TBMs in soft rock can exceed 100 ft per day. However, tunnel excavation is a systematic and industrial production process that can only proceed at the speed of the slowest element in the system. Thus, technological advances that increased the speed of rock fragmentation could not be fully applied until the muck conveying systems that remove excavated materials could be designed to keep pace. Advance rates for rock excavation also could be improved with more efficient methods of installing supports and reinforcements. Ground control measures to prevent loss of confinement and therefore face instability, chimneys,[4] cutterhead blockage, and tunnel collapse in blocky and highly fractured ground could be improved. Improved methods for predicting and mitigating spontaneous, energetic, and sometimes hazardous rock fracture (rock-bursts) in highly stressed competent rock could minimize risk and result in more repeatable construction practices, more consistent work products, and better control of construction times and costs. This parallels a need for fundamental advancements in our understanding of rock mechanics and rock behavior, including three-dimensional rock fracture mechanics.

With better quantification of rock and excavation tool interactions, there can be increased use of systems that sense their own progress and functioning (smart systems), more automation in underground construction (e.g., robotics) to reduce risk to workers, and enhanced ability to probe ahead of the excavated face to detect changes in material properties. Modern TBMs are computer-controlled, high-tech machines that use laser guidance systems and sensors to obtain real-time information on system performance. Data obtained during excavation can be used in feedback loops to optimally adjust, for example, thrust and rotation speeds.

Tunneling and Ground Support Technologies in Soil

Indurated rock may be more difficult to fragment while tunneling, but tunneling in soils (unconsolidated materials) presents challenges related to ground support, muck removal, and excavation-induced effects on the surrounding ground (e.g., ground settlement, lowering of the groundwater table). Improving control of deformation due to tunneling in soil is a key to limiting damage to existing infrastructure and improving capacity to sustainably develop underground space in most urban areas.

Tunnel lining is also an area where efficiencies can promote sustainability. Costs, for example, may be reduced when initial and permanent support (e.g.,

[4] Vertical openings from the tunnel to the surface created as a result of displacement of material into the tunnel.

construction support and final lining) are integrated, or when thicknesses of linings are reduced through use of new higher-strength materials. Other beneficial advances could include enhancing techniques to decrease overcut during mining, improving directional control during mining to decrease the allowances for alignment errors, and developing more efficient and rapidly deployed formwork for cast-in-place linings.

Tunnel Boring Machines in Soil

Recent technological developments for tunneling in soil include the earth pressure balanced TBM (see Figure 6.2) and the slurry shield TBM (Figure 6.3). These simultaneously excavate and support a tunnel face by creating a separate chamber (bulkhead) that closes off the face from the rest of the tunnel. These advanced technologies greatly mitigate risks of tunneling in saturated soils including those associated with running soils at the face, significant settlement in poor ground, and excessive water pressure. In the past, machine tunneling was not possible for deep tunnels under high water pressures without the use of compressed air in the tunnel—an expensive process with significant safety and risk issues. Comparatively, the earth pressure and slurry technologies greatly minimize settlement and associated damage, increase excavation speeds, and reduce costs. As a result, increasingly larger tunnels have become safe and economical to construct in poor soil conditions and allow such tunneling to contribute to more sustainable solutions for transportation and other infrastructure needs.

Sequential Excavation Methods

Sequential excavation methods (SEM) include multiple applications such as the New Austrian Tunneling Method and the Sprayed Concrete Lining Method that rely on integrating the excavated rock or soil into the tunnels' supporting structure (FHWA, 2009). SEM design includes deliberately relieving stress by mobilizing the ground around a tunnel to the maximum extent possible through controlled deformation. The methods involve commitment to both a design philosophy and a construction method largely dependent on the observational method. Initial primary support of the tunnel is designed with consideration for load-deformation characteristics appropriate to existing ground conditions; the required installation speed is based on the magnitude and speed at which ground deformations develop. The primary lining provides initial support during construction, preventing roof collapse. Instrumentation is installed to monitor deformations in the initial support system and to inform of changes needed in the support design and excavation sequence. The tunnel is sequentially excavated and supported, and the excavation sequences can be varied. The permanent support is usually (but not always) a cast-in-place concrete lining. SEM can be applied in both shallow and deep excavations, but in either case the effects of deformation on surrounding structures and the surface need to be considered.

Ground support in SEM can be selected or optimized in real time to match the ground conditions uncovered during excavation. This feature is intended to support the least favorable conditions encountered in the alignment and is in contrast to a pre-selected support system that does not accommodate variation in conditions along an alignment. The amount and the cost of materials needed for ground support can thus be reduced. However, the approach has its limitations: continuously changing the support scheme may reduce construction crew productivity and delay project construction. Further, the decision processes used in SEM are dependent on information that is not readily available (e.g., as related to ground condition, ground response, and movement). Research is needed in these areas to improve SEM application and contribution to more sustainable underground space development.

Trenchless Technologies

Many new underground utility construction and repair technologies have emerged in the past 40 years that allow installation, replacement, or repair of underground utilities or conduits without excavating a continuous trench from the surface. Although "trenchless" also applies to larger bored tunnels, the term typically refers to urban-utility-scale technologies rather than to rail, metro, or road tunnel installations. Trenchless technologies introduce new solutions for minimizing surface disruptions into short and long-term planning, design, and operation of underground systems. This will be especially true as the techniques become more commonly used.

Descriptions of trenchless technologies can be found in books and reports (e.g., Kramer et al., 1992; Thompson, 1993; Stein, 2002, 2005; Najafi and Gokhale, 2005), multiple conference proceedings (e.g., ISTT, 2010), journals (e.g., Broch et al., 1986), and magazines.[5,6] Figure 6.7 illustrates major trenchless technologies and applications. Because many technological advancements have centered on better equipment or new processes, complete understanding of the processes from a design and application perspective has lagged behind adoption in the field. Much information about the technologies is obtained directly from installation contractors, manufacturers, suppliers, or through trade associations. This potential lack of communication between engineering designers and construction contractors is problematic. In addition, the relatively new ability to install utilities at low cost beneath existing shallow utilities without major disruption to the utilities or traffic provides the ability—and even incentive—to deepen the existing spaghetti layout of utilities as pictured in Figure 1.6. However, this could interfere with future placement of major transportation or water and sewer infrastructure. In other words, such construction technologies heighten the need

[5] *Trenchless Technology*, magazine published by Benjamin Media, Peninsula, Ohio.
[6] *Underground Construction*, magazine published by Oildom Publishing, Houston, Texas.

FIGURE 6.7 Slurry shield machine for the East Side Access Project, Queens, New York. (left) Assembly of two tunneling boring machines (TBM) and their trailing gear in the launch pit. (right) Closeup of the cutter head of the TBM. SOURCE: Y. Hashash.

to address long-term planning and the ability to choose the most sustainable and best uses of urban underground space.

Ground Improvement Technologies

It is often necessary or productive to temporarily or permanently change soil or groundwater properties during underground construction to ease facility design, construction, or operation. For example, ground freezing can temporarily turn a weak, saturated soil into a solid and nearly impermeable material; dewatering can ease construction problems both in terms of water flow and soil stability; and grouting can be used to stiffen or change the permeability of soils. Ground modification techniques and the development of new materials used for geoenvironmental and geomechanical applications (e.g., bio- and nanotechnologies; NRC, 2006) are active areas of research and field application. Furthering application of established methods (e.g., jet grouting and compensation grouting), as well as developing new approaches and materials, offer the potential for cost reductions and performance improvement. However, as for trenchless technologies, application of new ground improvement technologies may outstrip theoretical understanding of the methods, meaning that application of these techniques in design may not be optimized. Further research and development is needed to improve existing methods, enhance understanding of their application, and allow better engineering with, for example, in situ biotechnologies, nanotechnologies, and other ground improvement and remediation techniques that can reduce the

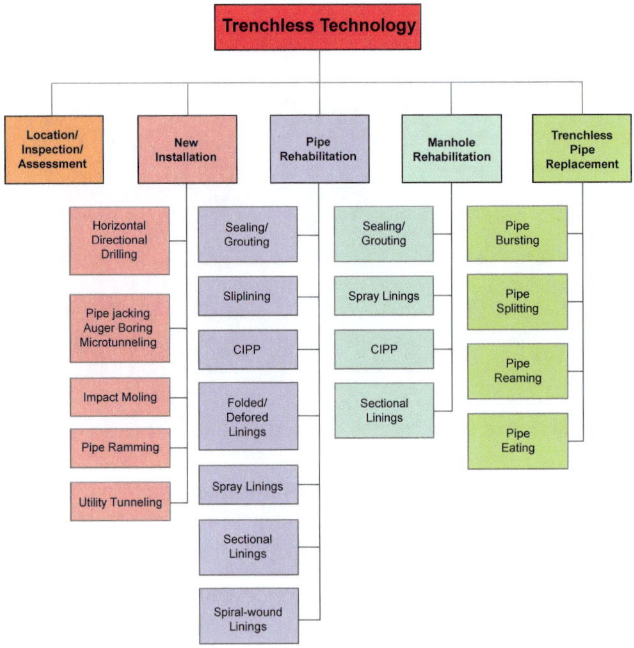

FIGURE 6.8 Major techniques grouped under "trenchless technologies." CIPP designates cured-in-place pipe. SOURCE: Raymond Sterling. Reprinted with permission from the author.

resource use and be gentler on the environment than can more traditional construction methods.

Monitoring During Construction

Geotechnical and construction monitoring programs provide the basis for understanding ground response to excavation and the impacts of construction on existing structures and the natural environment. Monitoring for construction may be more intensive than monitoring for infrastructure operation, but it is usually of shorter duration (often less than five years). Monitoring instruments may be exposed to harsh conditions during construction and need to be more robust. Monitoring the behavior of the underground facility and surroundings during construction helps to ensure safety, assess performance, validate design, and inform necessary design changes. Figure 6.8, for example, shows the level of detail possible using laser scanning to monitor excavation progress. Changes in the image over time provide information about work progress but also could provide information about the effects of excavation on, for example, nearby infrastructure. Monitoring water volumes and pressures and water and air quality

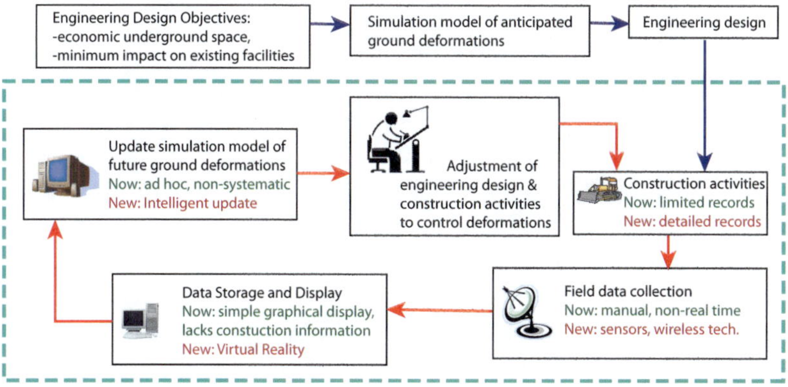

FIGURE 6.9 Monitoring data is continuously fed back into enhanced design and construction. SOURCE: Modified from Hashash and Finno, 2008. Used with permission from ASCE, ©2008.

helps to protect groundwater resources and air quality above and below the surface. Contractors typically monitor and analyze construction processes and operations (e.g., shift utilization, equipment downtime, repair or replacement, ground support installation, muck volumes, and grout takes) to identify opportunities to improve efficiencies.[7] Contractors also continuously monitor ground gases (e.g., methane, carbon monoxide, hydrogen sulfide) in excavations.

The use of sensors to measure geotechnical and structural displacements of individual structures, projects, or operations is well established in practice (for example, see Dunnicliff, 1993), and engineers continually improve their ability to use those technologies in ever more innovative ways. However, a major premise of this report is the need for a system of systems approach, which implies a need to create integrated systems that can monitor urban system conditions. Information regarding all elements of a system can be captured during all phases of their life cycles, allowing observation of the effects of changes observed in one system element on other system elements. Data from underground systems are expensive to collect as an isolated task, and hence it would be best to gather useful data whenever underground work provides such an opportunity. The past decade has seen a revolution in the developments of sensors, sensor technologies, and information technology infrastructure to allow integrated monitoring systems that provide timely feedback to designers. Figure 6.9 shows what such a feedback loop might look like.

Opportunities to advance construction monitoring include the means to collect and analyze detailed real-time monitoring data during construction

[7]This information is often considered the property of the contractor and is not necessarily shared with the owner.

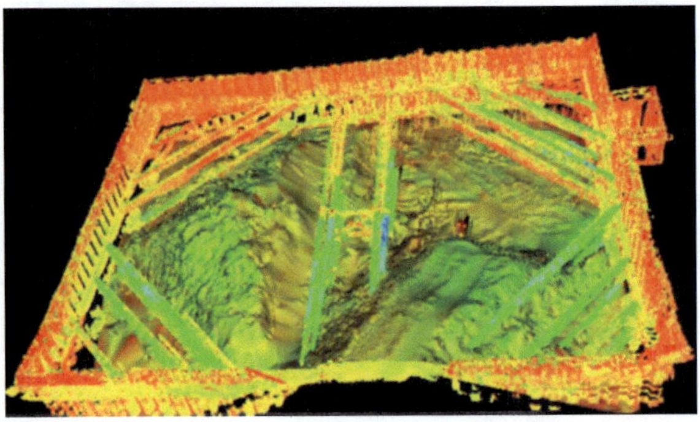

FIGURE 6.10 Example of laser scanning to monitor excavation progress. Image courtesy of Y. Hashash.

equipment operation. For example, cutterhead operation data (e.g., torque, power, thrust, penetration per revolution, disc cutter forces, rotational speed, bearing temperature, and disc location) increase knowledge of excavation processes, in particular geologic conditions. In a similar manner, continuous measurement of the volume, weight, and advance rate of removed spoil is an important indicator of ground displacement and settlement control (see Box 6.3 for example). Benefits of doing this have been recognized for some time, but the ability to apply—by practical means—the data gathered is still emerging (for example, see Box 6.4). Other opportunities come from advances in surface imaging and image processing. Surface displacements of an entire excavation face over time, for example, can be recorded using photographic, laser-based, or alternative imaging systems (see Figure 6.7). Some geologic features can be automatically identified from the scans. Such visual data could be integrated with conventional numerical method data to develop as-built plans and specifications for archiving purposes.

As a general trend, advances in construction monitoring are moving away from physical measurements at a limited number of points, to widely distributed, wirelessly connected sensor networks, and to areal scanning techniques. These data allow contractors to optimize construction processes and increase safety on the job (see Figure 6.10). Integrating data from multiple projects can facilitate improved equipment design, performance prediction, and understanding of zonation and geologic material properties. Monitoring for enhanced safety is also becoming more critical as larger diameter and shallower infrastructure (e.g., tunnels and stations) facilities are increasingly proposed for urban sites.

TECHNOLOGIES FOR EFFECTIVE ASSET MANAGEMENT

The ability to operate and maintain underground infrastructure over the long term is essential to its sustainability and to that of the larger system to which it belongs. All built systems require and need to accommodate maintenance,

BOX 6.3
Construction Monitoring in the Crossrail Rail System, United Kingdom

A system of 21 km of twin-tube rail tunnels is being constructed under London (United Kingdom) to expand existing network rail systems.[a] Extensive real-time data monitoring and centralized data management are critical parts of risk control and the processes to, for example, monitor construction progress and control of land movement, surface settlement, and volume loss (Reynolds, 2010a,b). For example, to control extraction and coordinate lining build rates, a measurement system determines spoil extraction weight on the fixed spoil conveyor that runs along the TBM, and video and laser scanning of spoil on the conveyor is monitored as a supplemental (but less reliable) weighing method (Reynolds, 2010b). The Underground Construction Information Management System (UCIMS) brings together all data from geotechnical and construction instrumentation and monitoring into a centralized resource. Contracts may require probe hole drilling during excavation from which drilling rates and water inflow volumes and pressures can be evaluated and geologic problems can be forecast. In some cases, geophysical probing ahead of the excavation face has been used to predict problematic conditions.

[a]See http://www.crossrail.co.uk/tunnelling/.

especially those operated over decades or centuries. Maintaining underground infrastructure is, in some ways, analogous to maintaining an aging fleet of aircraft through use of nondestructive methods of evaluation and repair. Sustainable infrastructure systems need to be designed to allow flexibility and to accommodate rapidly changing technologies, new safety and health standards, and changing cultural needs and demands—in other words, sustainability is dependent on the ability to accommodate technological and social evolution. For example, underground conduits can be reused if their size, condition, and alignment fit new needs. Low-pressure manufactured gas systems can be converted to high-pressure natural gas systems, for example, by slip lining new pipes within the old. Existing underground piping systems of various kinds can be (and have been) adapted to deliver fiber optic connections to homes and businesses.

In the shorter term, operation and maintenance of underground infrastructure represents a sizable annual investment. Many forms of underground infrastructure are difficult or costly to inspect. Most underground piping, for example, does not allow person entry for direct inspection, and taking pressure pipes (e.g., for gas, oil, or water) out of service for inspection entails major disruptions and revenue loss. Underground rail infrastructure (especially highly used metro systems) present special challenges related to access for inspection and rehabilitation. Compared to other forms of underground infrastructure, road tunnels are probably well maintained because the interiors are readily visible—problems may be identified more quickly. Much underground infrastructure, however, can be

BOX 6.4

Networked Instrumentation for the Transbay Transit Center, San Francisco, California

The Transbay Transit Center (TTC; see http://transbaycenter.org) is a $2 billion project to replace the existing Transbay terminal in downtown San Francisco with a modern regional transit hub that will connect 11 regional and city transit systems and will be the terminus point for the High Speed Rail from Los Angeles. The construction of the TTC requires a 60-ft-deep, 185-ft-wide, and 1,500-ft-long open-cut excavation in relatively soft soils adjacent to several high-rise and low-rise buildings including one of the tallest buildings in San Francisco.

The excavation is unprecedented in scale on the U.S. West Coast and includes an extensive monitoring program to help minimize disruption to surrounding facilities. The project employs state-of-the-art in-place sensors placed at depth in the underlying soils, as well as on surrounding buildings, utilities, and other infrastructure (see Figure 1). The sensors measure the response of the soils and infrastructure to construction activities. Cameras are used to continuously capture construction images. Microphones immediately point cameras in the direction of sources of sudden noise. Data for all sensors and imaging equipment are streamed wirelessly into cloud storage and accessed from anywhere in the world via a password-protected, web-based interface called the Global Analyzer (see Figures 2 and 3). The Global Analyzer synthesizes all data streams and issues automated alerts via email and mobile text messaging should any of the instruments exceed pre-determined thresholds. This information is complemented by a Twitter feed whereby contractor personnel and design team engineers post brief descriptions of construction activities—thus developing a record of construction that can be later referenced.

FIGURE 1 (Left) Placement of an in-place inclinometer with wireless networking. (Right) Automated total station with camera and dedicated wireless hotspot/node to continuously measure movements of targets placed on infrastructure around the site (Photos Courtesy of GeoInstruments).

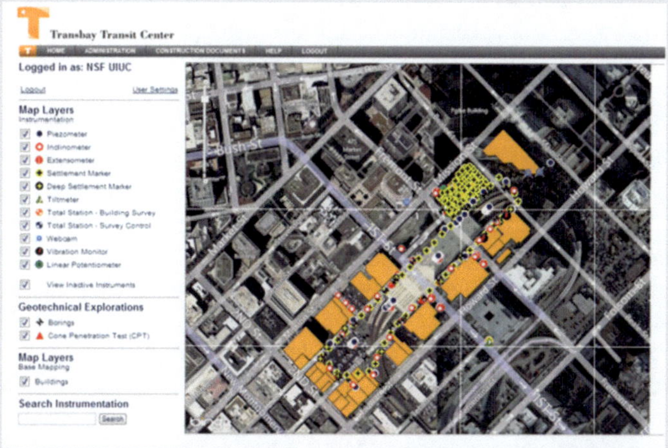

FIGURE 2 The Global Analyzer cloud-based sensor data and documentation interface for the TTC. Designed and implemented by Arup North America Ltd. for Transbay Joint Powers Authority, San Francisco, California.

FIGURE 3 TTC project wireless network for live data and construction image transmission. Automated total stations, shown in green are used as wireless nodes for dedicated project wireless network. Designed and implemented by Arup North America Ltd. for Transbay Joint Powers Authority, San Francisco, California.

under-maintained for decades. Many deterioration, maintenance, and operational issues in underground facilities stem from, for example, groundwater intrusion into facilities or groundwater-induced corrosion of the facility structure. New materials for sealing underground structures or self-sealing of leaks or cracks would help to keep groundwater out of structures, and innovative design can incorporate drainage water into the aesthetic or energy concepts for the facility. It is also worth noting that improving asset management is not a question of just collecting more data—but of collecting worthwhile and cost-effective data that are then effectively analyzed to glean the useful information that guides decisions on maintenance, rehabilitation, and replacement.

Inadequate maintenance has lately received public attention thanks to various reports on the status of America's infrastructure (e.g., ASCE, 2009). Increased attention to the problems caused by deteriorating infrastructure has resulted in some investment in improved technologies for inspection and rehabilitation. Much inspection work can now be done remotely using in-pipe robots, inspection "pigs," or even free-floating devices that pass through pipes "listening" for leaks or gas pockets and that can provide information about flow rates and properties of the pipes such as the locations of valves and pipe joints.[8] Remote fault detection through electrical and fiber optic cables can identify the position and nature of some defects. The concept of monitoring is taken even further with an increasing interest in "smart structures" (e.g., Wadhawan, 2007), in which wireless sensors are placed to monitor the structural health and long-term performance of various aspects of a structure. However, the challenges of use of sensors over decades remain formidable. The harsh environment in the underground can lead to their rapid deterioration in a relatively short period of time (e.g., Hoult et al., 2009; Stajano et al., 2010). Limited radio connectivity and power underground also can be a problem for wireless sensor networks (e.g., Bennett et al., 2010), and rapid technological developments in the wireless sensor field can render existing sensors nearly obsolete in less than a decade. Significant developments are needed in this area to fully achieve the vision of smart underground structures.

Addressing the immense maintenance deficit is vital for the sustainability of urban systems and requires a multifold approach. Significant incentives to invest in inspection, rehabilitation, and replacement of infrastructure, as well as to develop technologies that provide the cost-effective means to do so while minimizing disruption to city life and commerce, are necessary. New underground facilities must be designed and constructed with greater attention to lifecycle cost analyses (see Chapter 5) that account for modes of inspection and repair, sustainability, resilience, and end-of-useful-life considerations.

[8]One such device is manufactured by Pure Technologies Ltd. This device reportedly can be deployed in live pipes for up to 12 miles and detects leaks by emitting acoustic pulses transmitted to receivers attached to the pipe at known locations. The locations of leaks and air pockets are determined by analyzing the arrival times of the pulses at the receivers (Pure Technologies, 2012).

New technologies that reduce the costs of operation and maintenance of underground facilities, make them safer, and extend their useful lives could have a major impact on future costs and operational reliability of underground facilities. Basic research needs lie in better understanding of corrosion and other deterioration mechanisms both for in-use materials and for new materials and in the development of improved non-destructive testing approaches. Applied research needs include the continued improvement of inspection, assessment, rehabilitation, and replacement technologies—including ways to upgrade occupied underground facilities to meet current expectations of health, safety, and comfort. Better design and planning options for the reuse of existing urban infrastructure and creating multi-use options for the future (e.g., new design concepts for facilities that are easier to rehabilitate or retrofit for enhanced service life or repurposing) would also improve possibilities for sustainability.

Data Technologies

This section discusses broader issues of integrating underground system operations with the total infrastructural and social fabric of a region. When properly carried out, this integration can optimize facility use, ensure that operation and maintenance needs are met, and enhance delivery of societal expectations. For example, integrating underground and surface transportation systems can assure smooth flow of vehicles, occupants, and cargo for an entire region. Similarly, efficient wastewater conveyance tunnel systems operate in tandem with daily cycles of water use in a city. Situational awareness for development, operation, maintenance, and integrated management of urban systems is vital. Society is increasingly reliant on electronic, automated, remote, and networked sensor systems to monitor interactions and inform automated or human-in-the-loop decision systems; however, such data collection systems are not foolproof.

Varied sensor systems have been developed to measure a myriad of underground environmental and space functionality factors, all in an effort to inform decision making about underground operations and safety—for example in underground railroad systems (Bennett et al., 2010) and water distribution systems (Shinozuka et al., 2010). Factors such as air and water quality, noise level, traffic load, temperature, structural integrity of support systems, loose bolts in liners, corrosion of reinforced concrete, water seepage, traffic flow, control signals, and changes caused by extreme events (e.g., earthquake, explosion, fire) are among those that can be monitored. The ability to monitor such factors is vital to infrastructure sustainability and resilience, but current monitoring efforts cannot be applied widely enough to provide an adequate picture of the performance of a complete infrastructure system.

Because operational data transmission systems are considered "permanent," there are incentives to invest in robust systems. Although sensor and supporting information technology infrastructure will be designed to require relatively limited maintenance to the best of current ability, they also must be reliable and accurate, self-calibrating, small in size, reasonably priced, easy to operate and maintain, and upgraded at routine intervals. Sensor systems are expected to provide data around the clock for years and withstand dust, pollutants, moisture, and stray currents, and in the most useful case, provide immediate notification of failure. Sustainable systems require that failure of an individual sensor does not take down the entire system. Power to operate sensors must be obtained through replaceable power packs or remotely acquired when a measurement is taken (i.e., a passive sensor), or by "scavenging" energy from the sensor environment (e.g., local vibrations, fluid flow). Additionally, sensors need to be hardened against vandalism or accidental damage. The longevity of many sensor systems, however, is not known.

Data retrieval that allows real- or near real-time reception and data interpretation is important for operational decision making. Many of the same challenges that exist for transmitting construction data also exist for transmitting operational data. Hard-wired data transmission systems that require dedicated lines are reliable but can be costly (especially in long tunnels), and they are not suitable everywhere. The number of sensors employed is limited to how many sensors can be wired. The use of wireless data transmission has recently increased, driven by advances in wireless Internet protocol access, wireless local area networks (LANs), and the proliferation of cellular-based mobile phone services. Wireless data transmission avoids the cost of wiring, but data transmission deteriorates significantly in underground and confined spaces, especially in long tunnels. Wireless data transmission is vulnerable to security breaches that can compromise the system operation (Stajano et al., 2010). Location-based information (e.g., global positioning system [GPS] data) can provide locations of system elements needing repair, relay real-time information regarding conditions in underground space, and map locations of automated sensing and maintenance devices, but GPS and cellular signals are difficult to receive underground.

In addition to infrastructure operators, some underground infrastructure users (e.g., rail passengers) may rely heavily on location-based services and are accustomed to easy accessibility. Further development of location-based technologies that allow for seamless transition from aboveground to belowground may encourage underground use by those who do not want to lose that functionality. Further, real-time traveler information (e.g., arrival and departure information) on flat screens in small businesses within or near mass transit systems, or via text messaging and email alerts to mobile devices, could help to ease congestion in and around transit stations (Zhang et al., 2011).

Many other opportunities remain to develop new sensors and integrated systems for enhancing operation of underground transport systems. For example, technologies that employ security camera images for structural evaluation may prove beneficial. Selected trains or maintenance vehicles could be equipped with high speed cameras or laser scanners for periodic documentation of tunnel conditions and compared using image reasoning algorithms to evaluate changes in structures.

Linking Data and Asset Management and Analytical Capabilities

Continuous streams of numerical and visual data can inform day-to-day operations, maintenance, and predictions of longer-term infrastructure performance and operations programs. However, interpreting the large streams of numerical and visual image data in real or near-real time can easily overwhelm human operators. Important information requiring action may be missed. Methods and automated systems to interpret these data and report problems to an operator would enhance optimal operation and maintenance of underground systems. Enhanced data management technologies can aid understanding of the performance of underground infrastructure as part of the larger urban system and allow planners to anticipate interdependencies and interferences that affect functionality and quality of service. Data management technologies such as Building Information Modeling (BIM) processes (e.g., Smith and Tardif, 2009) may make it possible to evaluate in greater detail the impact of new construction on existing systems installations, to evaluate the impact of existing systems on constructability of a new project, and to design sensing systems tailored for new and rehabilitated systems as part of an integrated urban system of systems. However, although these methods are extremely important, the technologies employed may become dated, and budget limitations make necessary data updates and accessibility challenging. Some data may need to remain secure. Private-public collaboration may be necessary to link, analyze, manage, and access system-wide data.

Systematic, standardized documentation of case histories related to underground infrastructure could help to expand fundamental understanding of excavation and support processes. Indeed, case histories are an important way to learn about the underground because they benchmark the state-of-practice, and provide information that may validate or disprove assumptions and models. Archiving of data and records associated with site characterization for infrastructure development, design, operation and maintenance, rehabilitation, reuse, and decommissioning would allow improved future planning and management in a manner that promotes sustainability long after the data are collected.

Information Security

Preserving, maintaining, and protecting data integrity against neglect, vandalism, time, or technological obsolescence are serious issues that threaten sustainable management. Capturing subsurface information is difficult enough; properly cataloging and maintaining it over long periods (e.g., 50 to 100 years and longer) is a significant challenge. Electronically archived data can become obsolete within just a decade or two when technologies change and the media on which they are stored can no longer be accessed. On the other hand, paper hardcopies of data have survived for nearly a century for some tunnel projects, but only if properly cared for.

The security of data during transmission or storage in central computer systems is an increasingly serious concern. The data can be accessed, manipulated, and corrupted by unauthorized parties to the detriment of safe or smooth operation of underground facilities. Because sensor data inform decisions affecting, for example, life safety (e.g., traffic operations, ventilation), the well-being of underground infrastructure occupants depends on the secure and proper functioning of the system. The concern becomes more serious when sensors are used in automated feedback loops (such as traffic management or supervisory control and data acquisition [SCADA] systems). Data sabotage can immediately impact underground facility operations. The cause of the troubles may be hard to track down.

As use of networked sensing and automated decision making becomes more pervasive, there is a need to develop secure data networks and authentication mechanisms to prevent malicious or accidental data corruption and manipulation. The most hardened networks are still potentially vulnerable to malicious attacks, and the National Research Council (NRC) has published multiple reports on issues related to information technology security (e.g., NRC, 2010a,b). In 2007, the NRC developed a strategy for cyber security research and promoted categories of research that included limiting impacts of security compromise (e.g., the design of secure systems, evaluation of security), enabling accountability (e.g., attribution, remote authentication), promoting deployment of security designs (e.g., "usable security"), deterrence (e.g., legal policies and measures), and speculative, "out-of-the-box" approaches to security (NRC, 2007). Resilience needs to be built into sensor systems, including human-in-loop decision making for critical components to mitigate against corrupt data.

TECHNOLOGIES THAT PROMOTE SUSTAINABILITY AND RESILIENCE

This section draws attention to some key issues related to the sustainability and resilience impacts of underground facilities, and specifically to how technological developments could promote improvements in these areas. Many of these issues already are considered in some form in the design and operation of underground facilities, but they take on special importance when considered in light of overall community sustainability and resilience. Other issues, such as the understanding and control of highly interrelated systems of systems, represent new areas of study with great future importance. The interconnections and interdependencies between individual infrastructure systems and the overall functioning and well-being of the social community and systems need to be considered.

Materials

Improving the possibility of sustainability necessitates consideration of the economic use of materials. To be considered are the materials used, their resource availability, the processes needed to create construction-ready products, the long-term availability of the materials, the energy (and carbon footprint) implications of use, and long-term environmental impacts. For example, even commonly used materials such as sand and gravel may be in short supply because of a lack of regional availability or to urban development and planning decisions that render sand and gravel resources inaccessible. In terms of energy use (see next section), concrete, a significant element in most forms of underground construction, requires a high level of energy input for its creation (termed embodied energy). Some commonly used construction materials have been proven to be detrimental to the environment and public health (e.g., various types of volatile organic compounds used in pipes that can contaminate groundwater systems, and asbestos used in cement piping).

Excavated materials from some tunneling projects could prove to be a resource for nearby construction projects. Millions of cubic yards of material may need to be removed from an excavation. Some of this material could be a source of sand, gravel, and rock. Some of this material, however, may end up being classified as hazardous and therefore need special handling and disposal. Still, a large volume of material may be suitable for other construction uses, or may be part of the solution to other sustainability issues. Box 6.5 describes the case of the reuse of excavated materials from the Boston Central Artery/Tunnel project to help reclaim a solid waste facility and turn it into a park operated by the National Park Service. Disposal or reuse of excavated materials is a serious issue that warrants further attention.

More sustainable use of materials could mean choosing underground design and construction options that use smaller quantities of materials or materials with improved performance, or it could mean incorporating more waste or by-product materials derived from other applications into design (e.g., geopolymers made principally from waste flyash). The lifecycle costs and benefits, however, need to be factored into decisions. For example, integrating primary (support for construction) and permanent ground support systems may allow for the use of less construction materials, but may affect the efficiency of construction operations. Maximizing the ability of the ground to be part of the support system, or reusing excavated materials from within or near a project, would help to increase efficiency in material use or reuse. New lining and underground construction technologies are needed that reduce material use and improve long-term facility performance. More informed decision making requires making available better information about the sustainability aspects of construction materials (e.g., availability, embodied energy) to the the designers.

BOX 6.5

Reuse of Excavated Materials from the Boston Central Artery/Tunnel Project

Spectacle Island in Boston (Massachusetts) Harbor is the site of a municipal solid waste facility in use until 1959. From 1959 until 1993, the landfill remained an uncapped source of leaching into Boston Harbor. Uncontaminated excavation material from Boston's Central Artery/Tunnel (the Big Dig) project was used to stabilize slopes on the island and fill and cap the landfill to convert it to recreational use. Excavated material was transported in more than 4,400 barge loads to the island beginning in 1992 (Barnett and Chin, 1998) and was used to cap the landfill with a 2-foot clay cap (MassDOT, 2012). The cap created an impervious layer that would serve to keep precipitation from mixing with the wastes beneath and leaching into the harbor. Excavated fill was mixed with biosolids from several waste composting facilities in New England to create topsoil that was subsequently vegetated to keep the cap in place (NEBRA, 2012). Approximately 2,400 trees and 26,000 shrubs were planted on the fill (MassDOT, 2012). In 2006, a 114-acre park opened after 15 years of cleanup activities. The park is operated by the National Park Service and houses a visitor center, several miles of hiking trails, and a swimming beach (NPS, 2012).

Energy and Carbon

The cost, availability, security of supply, and climate impacts of energy use have received worldwide attention in recent years, and scientists and engineers have been working toward developing calculators of the energy embodied in a variety of infrastructure and geotechnical systems (e.g., Chester and Horvath, 2010; Hammond et al., 2011; Soga, 2011). Without such calculators, it is difficult to understand the true energy costs of underground infrastructure. The underground offers multiple options for promoting energy efficiency and ameliorating climate change that are described throughout the report, but there remain planning, design, construction, maintenance, and other sustainability challenges to be better understood or integrated into practice to maximize the energy savings of underground infrastructure. More efficient or alternative methods to excavation or concrete production—both energy-consuming processes—may result in greater energy efficiency during construction.

Underground space use requires significant quantities of energy for ventilation, temperature control, lighting, fire detection, and other systems throughout the life of underground facilities. Some advances allow greater efficiency, but higher installation costs could deter their adoption. Development of technologies and space configurations that increase the efficiency of these systems will benefit facility operators and society at large. Light fixtures that accept lower energy demand lamps have been designed and are being specified for some new tunnels and retrofits. Ventilation systems are designed to minimize smoke danger associated with a large-scale fire and therefore have much higher installed-energy demands than needed for everyday operation. Minimum requirements for periodic ventilation system testing result in regular spikes in energy use. However, standards could be reviewed and changed to determine if less frequent or different testing methods can assure safe operation and thus reduce energy consumption over the life of the system. New technologies and processes that increase energy efficiency, and development of new and smaller space configurations that reduce the use of energy resources, will benefit facility operators and society at large.

Underground facilities can be constructed and used to conserve energy and create systems that achieve ground heat exchange with subsurface geologic materials in the urban underground. Important research on underground heat transfer issues began in the 1970s and 1980s (e.g., Geery, 1982; Bloomquist, 1999). Ground-coupled heat exchange systems have since grown in popularity in Europe (Sanner et al., 2003). However, they have not been used extensively in urban communities, and their long-term efficiencies when used in close proximity with each other have not been evaluated. Investigation of thermal effects and long-term impacts on both underground climate and underground space usage is warranted.

The use of lower temperature geothermal resources can help to reduce net emissions of greenhouse gases through the use of ground source heat pumps or similar heat exchange systems for heating or cooling structures and potable and nonpotable water for residential use. Such systems exchange heat from the earth to a structure in the winter, and vice versa in the summer, and in some cases, can be incorporated directly into the foundations of infrastructure. Although energy savings can be significant (DOE, 2012), there are environmental implications to be explored, including the selection of refrigerants (e.g., Forsén, 2005), and the long-term effects of potential ground temperature changes on aquifers and groundwater flow, chemistry, biota, and on underground infrastructure itself. Other subsurface use may become restricted in some areas because of the presence of a "forest" of geothermal boreholes.

The issues of energy and resource production and energy-related waste storage are not investigated in this report, but they are relevant in the context of underground engineering and sustainable development. Extraction of conventional energy and resources from below the surface could be made more efficient with improved excavation and extraction rates, for example. Oil and gas production from deep wells, coal mining, uranium mining, and more recently gas production from deep shale formations are all part of the complex interaction of the underground with our energy future. There is also interest in the use of carbon sequestration and other waste disposal technologies at relatively shallow depth. These technologies often generate intense public policy discussions about actual or possible environmental impacts and the relative merits of pursuing different policies for energy conservation or energy production. Often missing are the critical data and analyses in the public domain that properly assign benefits and liabilities to the various options that can appropriately inform critical future options regarding energy and climate.

Nontraditional energy sources could be better and more efficiently exploited with advances in underground engineering technologies. Geothermally active regions can be exploited, for example, by drilling into hot rocks and using the naturally high temperatures to produce steam and electric power (Duffield and Sass, 2003). Known conventional geothermal resources have the potential to produce approximately 9,000 MWe, and an additional mean power potential of 30,000 MWe are estimated from undiscovered sources (USGS, 2008). An estimated additional 518,000 MWe could potentially be generated from nonconventional geothermal methods including engineered geothermal systems (EGS) (USGS, 2008). Issues such as corrosion from the highly corrosive groundwater in geothermal fields have to be addressed. Similarly, long-term operation and maintenance issues are not well understood for EGS systems, especially given the high pressures and flow rates expected of EGS wells. Reliable high-temperature submersible pumps suitable for EGS development, for example, have been identified as a technology gap (DOE, 2008). Another major impediment to geothermal resource use is proximity of resources to where the power is needed.

REFERENCES

ASCE (American Society of Civil Engineers). 2009. Report Card for America's Infrastructure. Reston, VA: American Society of Civil Engineers [online]. Available: http://www.infrastructurereportcard.org/report-cards (accessed July 3, 2012).

Barnett, C.J., and K. Chin. 1998. Soil contamination assessment and characterization in urban tunneling. Pp. 79-86 in North American Tunneling '98, L. Ozdemir, ed. Rotterdam, Netherlands: Balkema.

Barton, N.R., R. Lien, and J. Lunde. 1974. Engineering classification of rock masses for the design of tunnel support. Rock Mechanics and Rock Engineering. 6(4):189-236.

Bennett, P.J., K. Soga, I. Wassell, P. Fidler, K. Abe, Y. Kobayashi, and M. Vanicek. 2010. Wireless sensor networks for underground railway applications: Case studies in Prague and London. Smart Structures and Systems. 6(5-6):619-639.

Bickel, J.O., T.R. Kuesel, and E.H. King. 1996. Tunnel Engineering Handbook, 2nd Ed. New York: Chapman & Hall.

Bieniawski, Z.T. 1976. Rock mass classification in rock engineering. Pp. 96-106 in Exploration for Rock Engineering, Proceedings of the Symposium, November 1976, Johannesburg, Z.T. Bieniawski, ed. Cape Town: Balkema.

Bieniawski, Z.T. 1989. Engineering Rock Mass Classifications. New York: Wiley.

Bloomquist, R.G. 1999. Geothermal Heat Pumps Four Decades of Experience. GHC Bulletin (December):13-18 [online]. Available: http://geoheat.oit.edu/bulletin/bull20-4/art3.pdf (accessed June 4, 2012).

Broch, E., R. Sterling, J. Zhao, and C. Rogers, eds. 1986. Tunnelling and Underground Space Technology. Oxford, UK: Elsevier Science.

Chester, M., and A. Horvath. 2010. Life-cycle assessment of high speed rail: The case of California. Environmental Research Letters. 5(1):014003.

DOE (U.S. Department of Energy). 2008. An Evaluation of Enhanced Geothermal Systems Technology. U.S. Department of Energy, Energy Efficiency and Renewable Energy [online]. Available: http://www1.eere.energy.gov/ geothermal/pdfs/evaluation_ egs_tech_2008.pdf (accessed July 6, 2012).

DOE. 2012. Geothermal Heat Pumps [online]. Available: http://www1.eere.energy.gov/geothermal/heatpumps.html (accessed July 6, 2012).

Duffield, W.A., and J.H. Sass. 2003. Geothermal Energy—Clean Power from the Earth's Heat. USGS Circular 1249 [online]. Available: http://pubs.usgs.gov/circ/2004/c1249/ (accessed July 10, 2012).

Dunnicliff, J. 1993. Geotechnical Instrumentation for Monitoring Field Performance, 2nd Ed. New York: John Wilcy & Son.

ECTP (European Construction Technology Platform). 2005. Strategic Research Agenda for the European Underground Construction Sector, Draft, October 2005. European Construction Technology Platform [online]. Available: http://www.ectp.org/documentation/FA-Underground-Constructions_SRA-VISION2030-23Nov05.pdf (accessed May 2, 2012).

El Salam, M.E.A. 2002. Construction of underground works and tunnels in ancient Egypt. Tunnelling and Underground Space Technology. 17(3):295-304.

FHWA (Federal Highway Administration). 2009. Technical Manual for Design and Construction of Road Tunnels-Civil Elements. Publication No. FHWA-NHI-10-034. U.S. Department of Transportation, Federal Highway Administration [online]. Available: http://www.fhwa.dot.gov/bridge/tunnel/pubs/nhi09010/tunnel_manual.pdf (accessed July 10, 2012).

Forsén, M. 2005. Heat Pumps: Technology and Environmental Impact July 2005: Part 1. Swedish Heat Pump Association [online]. Available: http://ec.europa.eu/environment/ecolabel/about_ecolabel/reports/hp_tech_env_impact_aug2005.pdf (accessed July 6, 2012).

Geery, D. 1982. Solar Greenhouses: Underground. Blue Ridge Summit, PA: Tab Books.

Grimstad, E., and N. Barton. 1993. Updating the Q-System for NMT. Pp. 46-66 in Proceedings of International Symposium on Sprayed Concrete - Modern Use of Wet Mix Sprayed Concrete for Underground Support, Fagernes. Oslo: Norwegian Concrete Association (as cited in Palmstrom and Broch, 2006).

Hammond, G., C. Jones, F. Lowrie, and P. Tse. 2011. Embodied Carbon: The Inventory of Carbon and Energy. A BSRIA (Building Services Research and Information Association) Guide BG 10/2011. Bracknell: BSRIA.

Hashash, Y.M.A., and R.J. Finno. 2008. Development of new integrated tools for predicting, monitoring and controlling ground movement due to excavations. ASCE Practice Periodical on Structural Design and Construction 13(1):4-10.

Hashash, Y.M.A., Q. Fu, J. Ghaboussi, P.V. Lade, and C. Saucier. 2009. Inverse analysis- based interpretation of sand behavior from triaxial compression tests subjected to full end restraint. Canadian Geotechnical Journal. 46(7):768-791.

Hoek, E. 1994. Strength of rock and rock masses. ISRM News Journal 2(2):4-16.

Hoult, N., P.J. Bennet, I. Stoianov, P. Fidler, C. Maksimovic, C. Middleton, N. Graham, and K. Soga. 2009. Wireless sensor networks: Creating 'Smart Infrastructure." Proceedings of the Institute for Civil Engineering - Civil Engineering. 162(3):136-143.

IMWS (Institute for Mechanics of Materials and Structures). 2009. Artificial Ground Freezing as Temporary Support in Underground Excavation. Institute for Mechanics of Materials and Structures, Vienna, Austria [online]. Available: http://www.imws.tuwien.ac.at/en/research/research-projects/alte-projekte/artificial-ground-freezing.html (accessed July 23, 2012).

Institution of Civil Engineers (Great Britain). 1996. The observational method in geotechnical engineering. London New York, NY, Thomas Telford: American Society of Civil Engineers [U.S. distributor].

ISTT (International Society for Trenchless Technology). 2010. 28th International No-Dig Conference and Exhibition 2010, November 8-10, 2010, Singapore. Alexandria, VA: ISTT.

Konda, T. 2003. Reclaiming the underground space of large cities in Japan. Pp. 13-23 in (Re)Claiming the Underground Space, J. Saveur, ed. Lisse, The Netherlands: A.A. Balkema.

Kramer, S.R., W.J. McDonald, and J.C. Thompson. 1992. An Introduction to Trenchless Technology. New York: Van Nostrand Reinhold. 223 pp.

Maidl, B., L. Schmid, V. Ritz, and M. Herrenknecht. 2008. Hardrock Tunnel Boring Machines. Berlin: Ernst and Sohn.

MassDOT (Massachusetts Department of Transportation). 2012. The Big Dig: Facts and Figures [online]. Available: http://www.massdot.state.ma.us/highway/TheBigDig/FactsFigures.aspx (accessed July 5, 2012).

Muir Wood, A.M., A.M.M. Wood, and M.I. Brunel. 1994. The Thames Tunnel 1825-1843: Where shield tunneling began. Proceedings of the Institution of Civil Engineers – Civil Engineering. 102(3):130-139.

Najafi, M., and S.B. Gokhale. 2005. Trenchless Technology: Pipeline and Utility Design Construction and Renewal. New York: McGraw Hill.

NEBRA (New England Biosolids and Residuals Association). 2012. Putting Biosolids to Use: Spectacle Island, Boston Harbor [online]. Available: http://www.biosolids.com/Headlines/pdf/NEBRACaseStudySpectaclIsl01.pdf (accessed July 5, 2012).

Nicholson, D., C.M. Tse, and C. Penny. 1999. The Observational Method in Ground Engineering—Principles and Applications. Report 185, CIRIA, London.

NPS (National Park Service). 2012. Island Facts: Spectacle Island. Boston Harbor Islands National Recreation Area [online]. Available: http://www.nps.gov/boha/historyculture/facts-spec.htm (accessed July 5, 2012).

NRC (National Research Council). 2006. Geological and Geotechnical Engineering in the New Millennium. Washington, DC: The National Academies Press.

NRC. 2007. Toward a Safer and More Secure Cyberspace. Washington, DC: The National Academies Press.

NRC. 2010a. Toward Better Usability, Security, and Privacy of Information Technology: A Report of a Workshop. Washington, DC: The National Academies Press.

NRC. 2010b. Wireless Technology Prospects and Policy Options. Washington, DC: The National Academies Press.

NRC. 2012. Induced Seismicity Potential in Energy Technologies. Washington, DC: The National Academies Press.

Palmstrom, A., and E. Broch. 2006. Use and misuse of rock mass classification systems with particular reference to the Q-system. Tunnelling and Underground Space Technology. 21(6):575-593.

Parriaux, A., L. Tacher, V. Kaufmann, and P. Blunier. 2006. Underground Resources and Sustainable Development in Urban Areas. Paper No. 700. IAEG 2006 Engineering Geology for Tomorrow's Cities: 10th International Congress of the International Association for Engineering Geology and the Environment, September, 6-10, 2006, Nottingham [online]. Available: http://www.iaeg.info/iaeg2006/PAPERS/IAEG_700.PDF (accessed July 10, 2012).

Peck, R.B. 1969. Advantages and limitations of the observational method in applied soil mechanics. Géotechnique 19(1):171-187.

Pure Technologies. 2012. SmartBall® for Water and Wastewater Pipelines [online]. Available: http://www.puretechltd.com/products/smartball/smartball_leak_detection.shtml (accessed July 3, 2012).

Reeves, H. 2010. Attributed 3D Geologic Models in the UK and Their Application to Planning and the Sustainable Development of Underground Space. Presentation at the First Meeting on Underground Engineering for Sustainable Underground Development, June 1, 2010, Washington, DC.

Reynolds, P. 2010a. Monitoring Contract for Crossrail. TunnelTalk: Direct by Design, June 2010 [online]. Available: http://tunneltalk.com/Crossrail-Jun10-Monitoring-contract-award.php (accessed July 2, 2012).

Reynolds, P. 2010b. Settlement Control Measures. Crossrail Geotechnical Series 3: TunnelTalk: Direct by Design, August, 2010 [online]. Available: http://tunneltalk.com/Crossrail-Aug10-Settlement-control-measures.php (accessed July 2, 2012).

Sanner, B., C. Karytsas, D. Mendrinos, and L. Rybach. 2003. Current status of ground source heat pumps and underground thermal energy storage in Europe. Geothermics 32(4-6):579-588.

Schmidt, B.S., and Y.M.A. Hashash. 1998. Seismic rehabilitation of two immersed tube tunnels. Pp. 581-586 in Tunnels and Metropolises: Proceedings of the World Tunnel Congress '98, April 25-30, 1998, Sao Paulo, Brazil, A. Negro, and A.A. Ferreira, eds. Rotterdam: Balkema.

Sellers, E.J., M. Kotze, L. Dipenaar, and M. Ruest. 2010. Large scale concrete cube blasts for HSBM model. Pp. 389-398 in Rock Fragmentation by Blasting, J.A. Sanchidrian, ed. London: Taylor and Francis [online]. Available: http://www.aelminingservices.com/pdfs/whitepapers/AEL-white-paper-on-Rock-Fragmentation-by-Blasting.pdf (accessed July 10, 2012).

Shinozuka, M., D. Karmakar, P.H. Chou, S. Kim, H.R. Kim, and L. Fei. 2010. Non-invasive acceleration-based methodology for damage detection and assessment of water distribution system. Smart Structures and Systems. 6(5-6):545-559.

Skempton, A.W., and M.M. Chrimes. 1994. Thames Tunnel: Geology, site investigation and geotechnical problems. Geotechnique 44(2):191-216.

Smith, D.K., and M. Tardif. 2009. Building Information Modeling: A Strategic Guide for Architects, Engineers, and Real Estate Asset Managers. Hoboken, NJ: John Wiley & Sons.

Soga, K. 2011. Embodied energy and gas emissions of geotechnical infrastructure. Pp. 59- 74 in Geotechnics and Earthquake Geotechnics Towards Global Sustainability Vol. 15, S. Iai, ed. Dordrecht: Springer.

Stajano, F., N. Hoult, I. Wassell, P. Bennett, C. Middleton, and K. Soga. 2010. Smart bridges, smart tunnels: Transforming wireless sensor networks from research prototypes into robust engineering infrastructure. Ad Hoc Networks 8(8):872-888.

Stein, D. 2002. Technical Information System: Rehabilitation and Maintenance of Drains and Sewers. Berlin: Ernst & Sohn.

Stein, D. 2005. Trenchless Technology for Installation of Cables and Pipelines. Bochum: Stein & Partners.

Sterling, R.L., J. Anspach, E. Allouche, and J. Simicevic. 2009. Encouraging Innovation in Locating and Characterization Underground Utilities. Report S2-R01-RW. Transportation Research Board [online]. Available: http://onlinepubs.trb.org/onlinepubs/shrp2/shrp2_S2-R01-RW.pdf (accessed July 10, 2012).

Thompson, J. 1993. Pipejacking and Microtunnelling. London: Chapman & Hall. 227 pp.

USGS (U.S. Geological Survey). 2008. Assessment of Moderate- and High-Temperature Geothermal Resources of the United States. Fact Sheet 2008-3082 [online]. Available: http://pubs.usgs.gov/fs/2008/3082/pdf/fs2008-3082.pdf (accessed July 6, 2012).

Wadhawan, V.K. 2007. Smart Structures: Blurring the Distinction between the Living and the Nonliving. Monographs of the Physics and Chemistry of Materials No. 65. New York: Oxford University Press.

Zhang, L., J.Q. Li, K. Zhou, S. Datta Gupta, M. Li, W. Zhang, M.A. Miller, and J.A. Misener. 2011. Design and Implementation of Traveler Information Tool with Integrated Real-Time Transit Information and Multimodal Trip Planning. Paper #11-2383. Transportation Research Board Annual Meeting 2011[online]. Available: http://pubsindex.trb.org/view.aspx?id=1092445 (accessed August 1, 2011).

7

Institutional, Educational, Research, and Workforce Capacity

Addressing sustainability demands a certain level of societal commitment and capacity to do the necessary work. This chapter examines issues related to society's capacity to use underground engineering as part of the means to enhance urban sustainability. The committee's task included exploring advantages of underground development, identifying research to capitalize on underground engineering opportunities, suggesting a research track direction to enhance needed human capacity, and exploring the drivers for underground development that enhance sustainability. In deliberating its charge, the committee came to realize that current models for education, research, and practice in fields relevant to underground engineering are more likely to encourage ad hoc and independent activity rather than the interdisciplinary efforts that promote sustainability. Market forces in the United States often encourage needed workforce capacity growth and urban and infrastructure development, but advances are often driven by the need to solve a particular engineering challenge in a particular setting without necessarily considering the broader societal benefits and impacts. Current institutional systems are not designed to develop the kinds of capacity needed for sustainable development. A new framework is needed that will enhance societal capacity and the types of research, education, training, and practices needed for sustainable urban planning and infrastructure development.

Societal capacity is greater than workforce capacity and includes:

- Sufficient availability of appropriately trained and experienced engineers, planners, architects, technicians, and other professionals to teach, research, plan, design, construct, operate, and maintain effective and resilient underground facilities;

- An adequate government, university, and industry commitment to develop the research capacity needed to keep the United States at the forefront of science and technology developments related to urban underground construction and space use (including the mechanical and electrical systems that are part of underground infrastructure);
- Sufficiently well-informed citizens and decision makers who appreciate the long-term implications of underground space use on the quality of life in urban areas; and
- Adequate institutional planning, policy, educational, and research structures that support cross-disciplinary and cross-sector initiatives to optimize sustainability and resilience through the use of underground facilities.

The preceding chapters describe realized and potential contributions of underground infrastructure and engineering to a sustainable urban society, and many areas of research and action items are identified throughout. The committee was not asked to prioritize these items because to do so would require an assessment of greater complexity than this committee could have achieved given the scale of its assignment. Instead, the committee identifies common themes related to changes in approaches to urban planning and underground engineering education, research, and practice necessary to promote urban sustainability. In this chapter, the committee presents a series of observations, conclusions, action items, and research necessary to support the most productive use of underground engineering for sustainable urban development. The conclusions are largely focused on the institutional frameworks that would support societal capacity, without which sustainability goals are less likely to be obtained.

COORDINATED FORMAL PLANNING

Observation: There is little strategic coordination of underground infrastructure development in the United States.

Conclusion 1. Coordinated formal administrative support and management of underground infrastructure as part of an integrated, multi-dimensional, above- and belowground system of urban systems is vital to urban sustainability.

Potential actions:

a. Recognize responsibilities related to formal support for underground infrastructure as part of the total urban system through coordinated planning and operations, fostered technological development, and local and regional rule making.

b. Develop and encourage use of a system for consistent data collection, archiving, and access to be used by all facility owners and operators to aid decision making.

Research:

a. Explore within the federal government the most appropriate technical and administrative approaches to facilitate coordinated management of the underground as part of a total urban system. Recognize and coordinate with ongoing research in this area, for example, that conducted by the National Research Council (NRC) Transportation Research Board related to road projects.

b. Conduct a technology scan of how countries and cities around the world collect, manage, make available, and use three-dimensional geological and buried structure information.

Urban infrastructure generally, and underground infrastructure more specifically, is owned, constructed, operated, and maintained by many different private- and public-sector organizations to serve an even larger number of stakeholders. These different groups may each have their own unique missions, be driven by different goals, and have different financial vehicles, all of which may be divergent. Contractors hired to construct or operate underground infrastructure may not have long-term commitment to the infrastructure or the region. There may be little opportunity for owners and operators to understand the interdependencies between their respective infrastructure systems.

Consideration of the spatial and functional interdependencies of surface and underground infrastructure during all phases of infrastructure life cycle is vital to urban sustainability. However, cultural and political conventions in the United States tend to recognize, systematically plan, and organize only the real estate and air rights on or above the surface, effectively ignoring the valuable and nonrenewable real estate beneath our feet (with the exception of resource extraction). Further, since the 1980s, the United States has lacked a coordinated multi-agency federal thrust to keep U.S. research and technology at the forefront in underground development. Infrastructure development, in general, and underground infrastructure development, in particular, suffer in the United States from being organized by sectors and without any mission agency or other organization within the federal establishment dedicated to coordination across sectors. This coordination could lead to a better management of research investments and reduced risk for federal investments (particularly of large infrastructure projects), and could also be coordinated with investments by states and municipalities. Integrated, holistic, and three-dimensional planning is necessary.

All levels of government in many regions of the country are facing economic difficulties that may be the economic norm for years to come. The intergovernmental financial assistance system that has made many underground systems possible may not be able to invest in underground infrastructure as has been done in the past. Development of an institutional framework that catalyzes sustainable growth patterns through strategic targeted investments becomes even more important under such economic circumstances. Information management, information technologies, and communication will be key in facilitating the complex but efficient research and the design, construction, operation, and management of underground infrastructure.

Observation: Market forces in the United States encourage workforce capacity growth and urban and infrastructure development, but often in an ad hoc manner that may not be consistent with urban sustainability.

Conclusion 2. Development of underground space as part of sustainable urbanization requires expanded and coordinated communication with stakeholders to better incorporate site-specific conditions, greater flexibility, and long-term community needs into infrastructure system design and optimal lifecycle management.

Potential actions:

a. Establish a federally led interdisciplinary network or organization of organizations and institutions to guide sustainable patterns in underground infrastructure development and encourage interdisciplinary research and communication of findings among all disciplines and stakeholders. Stakeholders include, for example, designers, long-term planners, architects, safety specialists, and an array of engineering, geologic, geophysical, environmental, and contracting specialists from industry, government, and academia.

b. Develop mechanisms for integrated and holistic three-dimensional research and planning that include information management and communication technologies to facilitate complex research, design, construction, operation, and management of underground infrastructure.

Research:

a. Explore models for designing sustainability into engineered systems of urban systems that recognize interdependencies, vulnerabilities, complexity, and adaptability. Coordinate ongoing research in the United States and elsewhere on, for example, complex adaptive systems and human factors engineering (e.g., incorporating behavioral science, human performance and capacity, personnel and training, and human biology and physiology into engineered systems).

b. Develop conceptual models of the complex interactions among multiple systems (e.g., mechanical, human, and environmental) to improve understanding, reduce risk, and effectively manage infrastructure amid changing technologies, societal conditions, and expectations.

c. Research the behavior of those operating, maintaining, and using underground infrastructure during normal and worst-case operation scenarios to optimize the human-technical interfaces in a manner consistent with long-term values.

An institutional framework that catalyzes sustainable development and adequately revitalizes the U.S. educational and research capacity to address sustainable urban underground space is needed. Required technical human capital could be developed within this framework by bringing federal, state, and local agencies, the engineering and construction industries, and university educators and researchers under the same umbrella. Underground space development could then be addressed in a holistic manner through integrated educational and research programs that extend beyond traditional undergraduate, graduate, and continuing engineering education and training. This involves significant changes in the basic structure of several professional degree programs in the United States including planning, architecture, engineering, public administration, and social and economic policy—a difficult undertaking. The nation needs planners that understand underground space and economists that better understand how underground infrastructure supports lifeline service provision and a robust economic urban environment. The NRC Transportation Research Board and the National Earthquake Hazard Reduction Program might be studied to determine what elements of those organizational models might be incorporated into an institutional framework as discussed here. It is particularly important that engineers understand social and economic factors that contribute to urban sustainability, but it is just as important that other stakeholders involved in urban planning and underground development have realistic expectations of engineering.

Shared information on the relationships among individual systems and overall system performance is vital, and an ontology that is accepted across sectors and institutional cultures is needed for coordination and collaboration (see Box 7.1). Data and models used to understand the direct, indirect, and social costs of decisions related to individual infrastructure elements over the life cycle of the system can be the basis for better decision making related to, for example, performance versus needed investments for repair, rehabilitation, or replacement.

Observation: Complex ownership models for underground infrastructure confuse responsibility for routine inspections, maintenance, repairs, guidelines, budgets, and liability.

Conclusion 3. There is a need to understand the ownership and control models of underground space and to develop guidelines for funding and performing essential periodic inspections, maintenance, and repair of individual infrastructure elements.

Research:

a. Analyze multidisciplinary and holistic approaches to view the complex web of ownership, control, and responsibilities associated with maintenance and safety of underground infrastructure.

b. Examine multidisciplinary approaches to aid transition to more modern systems management.

Understanding ownership, liability, and responsibility for underground space becomes more important if infrastructure management is to support improved sustainability across the full complexity of interlinked underground systems. Safety related to failure of, for example, underground utilities also needs to be addressed. With the increasing appreciation of Supervisory Control and Data Acquisition (SCADA) systems and their vulnerabilities, anticipatory strategies need to be developed to investigate events and either direct threats to urban society, or those that are the result of cascading failures. Past underinvestment in infrastructure construction and rehabilitation increases current and future vulnerability as a result of inadequate inspections, unrepaired deterioration, inadequate system capacity, and lack of adaptation to new demands and challenges.

TECHNOLOGICAL LEADERSHIP

Observation: The United States was a world leader in many areas of underground science and technology when there was federal and industry investment in underground engineering research and development.

Conclusion 4. Maintaining global competitiveness in underground engineering education, technology development, and practice supports urban sustainability, resilience, and the standard of living of the United States.

Potential Action: Allocate resources for broader interdisciplinary education and technology development in underground design and construction.

Research: Expand U.S. research that advances and revolutionizes, for example, materials technologies, robotic construction technologies, laser guidance systems, geographic information systems, and enhanced computer analysis and visualization systems that improve the ability to model, design, plan, and reduce risk associated with complex underground systems (see Chapter 6 for more detail).

It can be argued that achieving and maintaining a technological leadership position in underground engineering is not necessary for the United States to reap all the benefits from effective urban underground facilities. The United States

BOX 7.1

Managing and Sharing Data

Poorly delineated interdependencies may represent emerging risks, particularly in relation to extreme events. For example, performance maintenance, protection from attack, long- or short-term costs, quality of service, or equity of access and supply need to be investigated in terms of space (area affected, geographical linkage to secondary impacts, etc.) and time (temporal evolution of impacts and recovery) in order to optimize design or operation. Different modeling tools will necessarily serve different sets of stakeholders, but the information developed by them will be most useful if their formats are compatible and if they have common spatial and temporal registrations. System data and models often need high security but the means need to be developed to share the relevant and necessary information for studies of interdependency with a targeted user community. Uncertainties about data pedigree often exist in many infrastructure databases, and protocols that can provide information on data quality, resolution, uncertainty, and trustworthiness for example are important. Likewise, the management and curation of massive real-time data flows from performance sensing arrays and smart systems will become more important, as will tools for data mining, protocols for metadata generation, and tools to support rapid data interpretation including visualization.

does benefit from technologies developed elsewhere, but it is not in the country's best interest to rely as strongly on imported technologies and expertise as is currently done. Many underground critical facilities are specifically designed to provide enhanced security and resilience in the face of potential extreme events or risks. Further, although outside the scope of this report, underground engineering is an important contributor to national defense and energy capacity. Reduced U.S. technological capacity in underground engineering can negatively contribute to economic growth and the global competitiveness of U.S. firms.

The United States has been a world leader in many areas of science and technology for underground construction (see Box 7.2) in the past. Partnerships with researchers at academic institutions in the past 40 years contributed to a continuous flow of ideas, enhanced understanding, and a high-quality graduating workforce that provided leadership for U.S. industry. However, that leadership is retiring, and few replacements have been trained. The majority of underground construction innovation (e.g., slurry walls, tieback anchors, micropiles, deep soil mixing, jet grouting, slurry and earth-pressure-balance tunneling machines, cured-in-place pipe relining systems, and many more) now comes from outside the United States.

Today in the United States, industry and research institutions continue to collaborate some on technology development, and research institutions often receive industry support for students and research. Much engineering, construction, and equipment manufacturing workforce knowledge, expertise, and training necessary for underground development occur through mentoring, on-the-job and

BOX 7.2
Once a World Leader

The United States has been a world leader in underground technologies in the past. For example, the first fully underground hydroelectric power plant was constructed at Snoqualmie Falls, Washington, in 1898 (PSE, 2009). Major developments in hard rock tunnel boring machines came about in the 1960s as a result of the decision of Chicago planners and the Metropolitan Water Reclamation District of Greater Chicago to build deep interceptor tunnels in the competent dolomite rock to eliminate sewage and storm water overflow into Lake Michigan (e.g., Hapgood, 2004). These projects engaged university researchers and resulted in knowledge growth. In the 1970s, there was intensive effort to improve underground construction technology as agencies recognized the growing need for underground space use in urban areas, particularly in conjunction with subway (with funding from the Urban Mass Transit Administration [UMTA]) and combined sewer and water projects (mandated by the U.S. Environmental Protection Agency [EPA]). These projects resulted in U.S. leadership in ground support technologies (e.g., rock bolting and tunnel lining) and tunnel boring machine design, invention, and manufacturing. With support from federal agencies including the National Science Foundation's Research Applied to National Needs (RANN) program, UMTA, the Department of Defense, EPA, and the Department of Energy, the United States made significant advances in underground construction technologies in the 1970s and 1980s. Additionally, innovations in the pipeline construction and utility industries created radical new possibilities for pipeline and utility installations through new concepts in trenchless excavation and the adaptation of directional oil well drilling technologies to cable and pipeline installations in the 1980s and 1990s.

project-specific problem solving, extensive use of overseas construction firms on projects, and collaboration with international engineers on temporary assignments. To remain competitive, firms such as Parsons Brinkerhoff have career development programs to make up for the smaller number of colleges and universities that provide hands-on underground engineering knowledge. Industry groups such as the North American Society for Trenchless Technology (see http://nastt.org/training) also provide courses for professionals on targeted topics. However, this training is not viewed—even by those with extensive industry experience on the committee—as a broad education, and there is minimal contribution from higher education institutions to these efforts.

There are advantages but also important limitations associated with industry-based training. Economic competiveness within industry means that knowledge gained by a specific firm tends to remain with that firm and may even leave the United States if the firm returns overseas at project completion. Commercial constraints may prevent industry from embracing the challenges associated with an integrated and holistic approach to urban development, as well as those associated with infrastructure sustainability and long-term performance. In contrast, advancements made at multidisciplinary research institutes are more likely to

BOX 7.3
Multidisciplinary Research Aiding Domestic Competitiveness

From 1977 to 1995, at a time when the United States was a world leader in underground engineering technologies and innovation, the research organization with perhaps the broadest mission related to underground construction was the state-funded Underground Space Center at the University of Minnesota. The center assembled a multi-disciplinary team to look broadly at issues affecting underground space use, including public policy, planning, architectural design, geotechnical engineering, and underground heat transfer, and it became a model for several other centers around the world that guide underground space use in their respective countries. These include centers at the University of Delft in the Netherlands, Tongji University, Chongqing University and Nanjing Engineering Institute and other universities in China, and the Urban Underground Space Center of Japan. While University of Minnesota center was successful in terms of research activity and maintaining its broad mandate, the lack of a stable base funding for its mission left it vulnerable to a university- and state-funding recession that resulted in its closure in 1995.

be of greater societal benefit while also resulting in a more educated domestic workforce (see Box 7.3). This is even more important as the country prepares to address projected urban, demographic, and climate-related challenges.

AN EDUCATIONAL FRAMEWORK

Observation: Lack of funding continuity that allows meaningful investment in equipment and faculty has resulted in a substantial reduction in the number of U.S. university programs dedicated to integrated underground engineering research and education.

Conclusion 5. There is a critical shortage in educational, training, and research opportunities for engineers who wish to learn and practice underground engineering in the United States.

Potential actions:

a. Develop national multidisciplinary, multi-institutional, cross-sector research centers that focus on different areas in underground engineering and sustainable urban infrastructure to produce the next generation of leaders in underground engineering.

b. Integrate graduate underground engineering studies with research programs or a critical mass of coordinated faculty activity to anchor research to existing programs. Create opportunities to specialize in particular aspects of underground engineering, but with a multidisciplinary approach.

c. Develop university consortia to aggregate faculty expertise; strengthen industry-university faculty relationships.

d. Teach better facility planning and management with a multidisciplinary approach through traditional, distance, or hybrid-style education formats. Traineeships (e.g., NSF's Integrative Graduate Education and Research Traineeships) could help to fund programs.

e. Expose undergraduates to multiple disciplines, issues, challenges, and opportunities associated with sustainable underground space use and engineering.

f. Develop continuing education opportunities for professionals.

g. Develop appropriate credentialing for inspectors, technicians, and operators of complex underground facilities.

Good engineering depends on strong analytical skills, creativity, ingenuity, professionalism, and leadership (NAE, 2004) as well as on accumulated knowledge based on old and new successes in underground works. Undergraduate programs that contribute to the kinds of knowledge discussed in this report include but are not limited to mechanical, electrical, civil, structural, geotechnical and geological engineering, planning, architecture, public policy, fire safety, and information technology. However, traditional programs in these areas do not prepare students for an integrated approach to practice. Some interdisciplinary programs in underground engineering at the graduate level conform to the American Society of Civil Engineer's Policy 465 to support a Master of Science (MS) degree (or equivalent) as prerequisite for professional practice (ASCE, 2007). Some examples of programs include the MS in infrastructure engineering at the University of California at Berkeley, the MS in sustainable and resilient infrastructure systems at the University of Illinois at Urbana-Champaign, and the infrastructure focus of the civil engineering program for the MS in Engineering at Louisiana Tech University (Brierley and Hawks, 2010). Graduate education at some schools includes specifically identified foci (e.g., the certificate in tunneling at the University of Texas at Austin) or specialization within a more generally named graduate degree program. Cooperative education and internships for all forms of education are especially important in underground engineering, which is less codified than, for example, engineering for structural building design.

Education and training has been integrated in some underground engineering programs including, for example, the tunneling and underground engineering group at the University of Illinois (1970s and 1980s), the Underground Space Center at the University of Minnesota (1977-1995), and the Trenchless Technology Center at Louisiana Tech University. Such research groups significantly influenced general practice and specific applications, but the size of the programs paled in comparison with the scale warranted by the level of national investments in underground space use and infrastructure.

Today, there is little expectation of the funding continuity that allows meaningful investment in equipment and faculty needed to support enduring and integrated research programs and the type of integrated graduate studies suggested here. This relates to the lack of continuous government focus on infrastructure issues in general, and underground infrastructure in particular. Relatively few university faculty in the United States engage in tunneling research, and many of those focus on tunnel performance in seismic and other extreme situations rather than on improving tunnel design and construction performance. The number of U.S. university programs dedicated to mining engineering has also reduced substantially since the 1960s: fewer than 20 exist today.

The decline of research in underground construction and tunneling in universities in the United States mirrors the fragmentation of U.S. government-sponsored activities in underground development research. An underground engineering workforce that supports sustainability cannot be created by simply merging educational programs of similar skill sets. This is true for several disciplines that are at the core of underground engineering such as geotechnical and mining engineering. Geotechnical engineering, for example, is often treated as a subdiscipline within civil engineering and hence competes for resources with structural, transportation, environmental, and other engineering disciplines. The number of geotechnical engineering faculty at a university may be only 1 or 2, and seldom more than 5 or 6 in even large civil engineering faculties of 30-40 professors.

Mining engineering education and training has suffered in part as a result of a reduction in U.S. mining activity in favor of overseas mine development. The loss of mining engineering programs, faculty, and students, given their similar core knowledge as their civil engineering colleagues, compounds the human capacity issues for underground engineering. Specialized knowledge areas such as tunneling have been put under pressure by state-mandated reductions in credit hour requirements for undergraduate degrees, the lack of interest by U.S. students in pursuing advanced degrees, and the limited or sporadic nature of funding opportunities for research in these fields.

IMPROVING PERFORMANCE

Observation: The complexity of urban infrastructure systems and uncertainties associated with system design and performance increase with greater and more varied demands on both above- and belowground infrastructure.

Conclusion 6. Engineers and urban planners could better improve whole lifecycle facility performance and overall urban sustainability with documented and validated risk-informed approaches to project planning and design that balance lifecycle project needs in terms of service delivery, ini-

tial costs, resilience against extreme events, and effective maintenance and operations.

Research:

a. Advance existing and develop new technologies for modeling uncertainty during all phases of infrastructure life cycle. These include invasive and noninvasive technologies for geologic site characterization (including existing and legacy infrastructure and materials); analytical and computational design methods; excavation, ground support, and monitoring technologies; and technologies for asset management including related to the management of data and security (see Chapters 6 and 7 for more details).

b. Develop strategies to investigate potential hazards, impending problems, and cascading evolution of problems, especially given current underinvestment in infrastructure system rehabilitation.

c. Engineers and planners could use extreme events to understand complex systems behaviors and interdependencies and to validate computational models of system performance.

Sustainability and resilience have only been considered in broad terms for a decade or two, and there are more questions than answers regarding what sustainability and resilience strategies are most effective. However limited our current knowledge is, however, it is necessary to act on the best knowledge we have while rapidly improving our grasp of the complex system interactions, and while developing metrics to assess progress. In this regard, the educational framework discussed above can create a new generation of professionals able to integrate technical disciplines with the emerging understanding of sustainability and resilience and integrate risk-informed approaches to design, construction, and management.

Large and complex underground facilities and networks represent major financial investments, provide critical functions and services for urban living, and must not degrade health and safety. For most cities, however, major underground projects are not a normal undertaking and hence present major challenges to policy makers and the professionals from the planning, architecture, engineering, financial, insurance, building code, and health and safety sectors that will be involved with such projects. Trusted information about alternatives, costs, benefits, and risks that can be used by all from those contributing disciplines is needed as are the means to improve that information as additional knowledge and experience is gained. Interdisciplinary research, education, and training that allow development of practical methods to determine, for example, the remaining useful life of utilities and services are needed. Consideration of topics such as how best to reuse or reconfigure underground space as technologies change are also part of performance and total lifecycle planning.

ADVANCING TECHNOLOGY FOR SUSTAINABILITY

Observation: Aging underground infrastructure may be susceptible to deterioration and issues associated with changing technologies, changing climate, and societal needs.

Conclusion 7. Underground space development requires a long-term commitment to technological advancements in an environment that is friendly to improved planning, innovation, and implementation.

Potential actions:

a. Design infrastructure that allows ease of access for inspections, maintenance, repairs, upgrades, and reconfigurations in response to new needs or technologies that allow such work to be completed at lower costs.

b. Consider resource needs, availabilities, and access when making administrative and technical decisions concerning development. These include energy resources (e.g., oil, gas, and other energy resources), industrial minerals, high-value or critical strategic minerals (e.g., gold, uranium, rare earth elements), and construction materials (e.g., gravel, sand, building stone).

c. Use appropriate models that demonstrate multiple potential scenarios and allow better infrastructural system planning based on local conditions.

Research:

a. Academia and system stakeholders could collaboratively develop long-term performance simulation models for complex systems and validate the results over time to understand dynamic responses and emerging system behaviors.

b. Explore how technologies and innovations from other industries (e.g., exploration tools, in situ analytical techniques, measurement-while-drilling systems, laser scanning, fusion of multi-sensor data) and civilian application of military research could be applied to underground engineering.

c. Conduct long-term research on the effects of the underground infrastructure on the natural and built environments to increase the capacity of decision making for society's best long-term interests.

d. Research comprehensively and on a common risk-cost-reward basis the long-term effects on sustainability of underground storage or disposal of urban wastes (e.g., municipal, sewage, or energy-related products).

Improved technologies can enhance the ability to select the most sustainable approach to underground space use by making such use cheaper or better. For example, the development of better planning, design, and construction technologies can reduce construction costs, minimize deterioration, increase resilience, and address geologic, hydrologic, environmental, thermal, and social issues that exist or may arise over time. Technologies have improved in the past decades, but, interestingly, many of the general areas in which improvements are regularly cited as being needed have not changed. For example, in 1989, the National Research Council identified the ways in which geotechnology impacts the U.S. economy, the environment, and national security (NRC, 1989). Multiple research themes deserving special attention were identified that could contribute to infrastructure development and rehabilitation including:

- influences of construction on nearby structures;
- trenchless construction technologies for installing and rehabilitating utility pipe networks (see Box 7.4);
- development and use of new materials such as plastic pipe, polymers, and geosynthetic materials to address infrastructure system needs;
- maintenance and renewal of aging infrastructure systems, including remote sensing systems to locate and assess infrastructure system quality; and
- an interdisciplinary approach to solving the diverse needs of complex infrastructure systems.

Research in many of these areas has improved U.S. capacity to develop underground systems, but research in these same areas is still warranted today, especially given national interest in sustainability and resilience. Chapter 6 provides a detailed discussion on needed technology innovations associated with site characterization, and underground infrastructure design, construction, operation, monitoring, and maintenance that could contribute to sustainable development.

Some specific technology development challenges and opportunities for research that would aid a more holistic approach to integrated urban system design and operation are highlighted in previous chapters and in Boxes 7.4 and 7.5.

LIFECYCLE APPROACHES

Observation: Few data exist regarding the environmental and social impacts and lifecycle sustainability of urban development that can inform technology and administrative decisions related to long-term (decades to centuries) infrastructure operation, maintenance, and reduced costs.

Conclusion 8. Comprehensive and scientific retrospective studies of the direct and indirect costs and impacts of various types of underground projects are needed to evaluate usefulness and economic, environmental, and social impacts so that future planning can maximize sustainability.

BOX 7.4

Specific Challenges for Pipe and Cable Systems

• Piping systems in the United States have expected service lives of 50 to 100 years, and cables have expected service lives of 10 to 15 years (EPA, 2008). Many systems in the United States have exceeded their expected service lives and may fail in coming years if not renovated or replaced.

• The development of new pipe and cable materials that perform better over longer life cycles, as well as of new smart underground infrastructure networks that monitor their own performance and condition are needed. Smart systems could allow improved prediction of needed repairs before costly failures occur. The result could be more intelligent infrastructure maintenance planning and integrated decision making. For example, needed repairs in an area could be coordinated, minimizing combined repair costs and closure of public rights of way.

• Three-dimensional position and performance information is important, especially given the premium now placed on new techniques to rehabilitate conduits and increase capacity of existing pipe in situ rather than creating new alignments.

• Utilidors that combine utility systems into compact and maintainable configurations may be effectively justified through development of workable scenarios for secure multi-utility facilities, lifecycle cost-benefit analyses, and effective transitioning strategies combined with demonstration projects.

• Future design standards need to include consideration of the role of individual system elements in the larger urban system over their life cycles. Standards also need to anticipate, for example, the effects of climate change in a region (e.g., drainage systems may require greater capacity to accommodate increased intensity, duration, and frequency of storms).

• Planning and design will need to accommodate multi-hazard approaches to risk-based management over the life cycle of systems and will need to consider long-term robustness, resilience, and sustainability during design and operation. For example, the impacts on groundwater resources and structural adequacy, buoyancy, water tightness, and corrosion will require increased attention in areas affected by changing groundwater levels (especially if coupled with saltwater intrusion).

Research:

a. Conduct comprehensive and scientific investigations to retrospectively identify the lifecycle performance of various types of underground infrastructure and to identify the aspects of project planning, design, construction, and operation that contribute most to project costs and performance. For example, track financial (both direct and indirect), environmental (e.g., air and water quality), and social impacts over an extended period (e.g., decades) following a project such as Boston's Central Artery alignment.

b. Develop common metrics for assessing sustainable development more generally, and for assessing specific economic, environmental, and social impacts.

c. Develop quantitative methods to compare the value of underground space

BOX 7.5

Mapping and Data Capture and Assessment Technologies

Urban underground space can be better managed with less labor-intensive means to map accurate positions of all underground utilities, perform essential lifeline service inspections, and manage the resulting massive databases. Reliable documentation of all things underground in an accessible and searchable database system would improve the ability of planners to maximize underground space use while minimizing construction and maintenance costs. The technologies also could lead to better long-term systems approaches to planning, construction, operations, and maintenance. The means also could be developed to dynamically link ground information models with feedback from construction or monitoring equipment to enable real-time characterization, response prediction, and decision making related to processes throughout infrastructure life cycles. However, underground data collection and transmission related to wired system robustness and wireless system transmission capabilities and energy requirements still present challenges. Sensors and network systems are needed that can be placed underground in widely distributed and self-organizing networks, that allow long-term operation (including calibration and location registration and configuration), and that can be operated remotely. Coordinated technology developments could be considered in areas such as low-power sensing and systems, power scavenging and harvesting, or the development of wireless signal transmission systems in the underground.

on a par with other urban resources (e.g., linked to market value of surface property) and in consideration of the impacts on future underground use (e.g., infrastructure may need to be placed in increasingly difficult ground conditions).

d. Compile data about sustainability aspects of various construction methods and materials (e.g., the availability of materials and energy embodied in production of materials).

Lifecycle analysis is a strategic tool that can inform decisions related to operations, maintenance, costs, and environmental impacts that affect sustainability. Understanding whether, for example, urban underground development precludes good stewardship of underground water resources in a region may require quantifying the amount of evapotranspiration, groundwater recharge, flow patterns, and pollution, among other factors, enabled because of different construction techniques or the preservation of natural landscapes. Retrospective analyses inform strategic prospective lifecycle cost analyses that, ideally, become part of local and regional planning processes. Decision makers that understand the true costs of infrastructure options over time are likely better poised to make decisions that support sustainability. Design decisions that affect sustainability include, for example, those that integrate initial (during construction) and permanent ground support systems that require less construction materials, use materials with improved performance, or incorporate more waste or by-product materials from other applications (e.g., concrete made from coal fly ash-based geopolymers).

Capturing all costs, such as diminished air quality or those associated with disruption and business losses during street closures, in a comparison of project alternatives remains a challenge and a topic for future research. Although costs may differ significantly among similar projects, discrepancies observed in the cost of a lane-kilometer of roadway in various countries, for example, suggest that investigating the detailed reasons for lower costs in some countries as compared to others would be worthwhile. Understanding such relationships would assist development of a realistic management framework that objectively distributes total costs over infrastructure life cycle. An entire infrastructure management framework could be informed that includes planning, documenting existing conditions, establishing land use requirements (both above and below ground), and issuance of permits for approved underground use (as directed by informed policy).

USER SAFETY AND COMFORT

Observation: Underground infrastructure can safely enhance the lives of millions, but few federal-level safety regulations exist to guide operational safety at a time when underground system complexity is increasing.

Conclusion 9. Greater user acceptance and occupancy of underground infrastructure and facilities are likely if underground spaces are planned with more consideration of utility, ease of access, wayfinding, safety, and aesthetics.

Potential actions:

a. Develop and adopt performance-based safety mechanisms and codes that not only account for today's underground occupancies (e.g., mixed use, multi level) and risks, but also allow for expansion and change of use. The International Code Council technical requirements, applicable National Fire Protection Association standards, and other related standards and guidelines could be expanded and made applicable to underground facilities.

b. Incorporate human factor and complex systems engineering concepts to guide threat recognition and technical and operational decision making for normal operations and for operations during times of stress (e.g., in response to extreme events).

c. Incorporate behavioral science, training, biology and physiology, human performance and capacity into safety codes and design.

Research:

a. Research the state of practice and best practices related to safety systems (e.g., hazard detection, notification, ventilation, fire suppression, emergency egress, and system integration). Develop appropriate minimum safety system requirements to incorporate into national-level guidelines and standards.

b. Compare international underground safety codes and guidelines with those applicable in the United States to identify inadequacies and guide future practice, recognizing existing efforts in this area (e.g., by the Federal Highway Administration).

Public acceptance and use of underground space will increase if underground infrastructure is more convenient and comfortable to use. One design challenge is long-range planning that incorporates strong connectivity within underground systems and with surface systems. This means creating usable reasonably connected underground systems that limit pedestrian travel time and lengthy vertical movement by stairs, escalators, or elevators. However, existing building codes may not be flexible enough to accommodate the types of design that increase convenience.

Building codes exist to protect the health and safety of those constructing, operating, or using infrastructure, but their slowly evolving nature leaves little room to benefit from evolving technologies. Further, existing safety codes, regulations, and standards designed to address known risks above ground are often inadequate for large-scale, sustainable development of the underground. Large-scale public use will require development of new and updated safety regulations that specifically address risk of the underground and activities (occupancies) therein.

Allowing variation in design based on better understanding of how to create safe but interesting and enjoyable underground space without greatly increasing costs and space requirements remains a challenge. Incorporating more human factors engineering into underground and urban system design and operation may improve the underground for safety, productivity, and aesthetics. Research into new materials and their behaviors, combined with risk assessments and management activities that incorporate, for example, provisions for emergency evacuations, rescue, and recovery would benefit the underground environment during normal operations, as well as during and following stressful events. Identifying and countering negative perceptions can be as important as safety and technical challenges and require their own research focus.

FINAL THOUGHTS

Observation: Underground space is a valuable but decidedly nonrenewable resource.

Conclusion 10. Underground space can enhance urban sustainability only if the underground is thoroughly understood and if underground use and reuse and the protection of the natural and built environments are incorporated into long-term total urban infrastructure system planning.

Potential actions:

a. Institute planning of all underground space as part of an evolving urban system to be carefully engineered or preserved for optimal long-term use and regional sustainability.

b. Establish reasonably intensive groundwater, soil, and infrastructure monitoring practices to track the health of the underground urban environment according to the general geologic conditions and use. Use data generated from a range of environments and situations to inform urban planning in other areas.

It is easy to look at a photograph of a city and envision a three-dimensional model of its surface structures, skyscrapers, and raised highways. This report challenges many urban planners, designers, engineers, researchers, contractors, and infrastructure operators to include the subsurface in this three-dimensional model, and to coherently link infrastructure between the surface and subsurface. Just as there is only so much surface area in a given city, there is only so much usable underground volume beneath the surface. However, unlike infrastructure on the surface, underground infrastructure cannot be easily removed or rebuilt when its useful life ends. Once subsurface geologic materials are removed and infrastructure elements or waste are put in their place, the subsurface cannot be restored to its original state and possibly may not be used for other purposes. For this reason, urban sustainability is dependent on thorough understanding of the underground and how best to plan for the use, reuse, and protection of underground resources—whether referring to natural energy or material resources, or to the underground space itself.

People have exploited underground space and resources for thousands of years to advance and protect survival, economic prospects, mythological culture, and spiritual growth. These endeavors involved high risks offset by the belief that the benefits of the underground exceeded the dangers—long before there was detailed understanding of the underground environment or sophisticated tools with which to explore it. However, early successes and failures in the underground helped build the substantial knowledge base that exists today throughout the world. The challenge now is to create a comparable legacy to sustain the nation's natural resources, economic efficiency, and social solidarity for the long term. This means expanding our knowledge base in ways that align our technical tools, collective perceptions, public policies, regulations, and procedures so that we can reduce risks to negligible levels, create needed services and spaces that function reliably and lift our spirits, and ultimately provide an integral and balanced support system for livable and sustainable urban areas.

REFERENCES

ASCE (American Society of Civil Engineers). 2007. Policy Statement 465 - Academic Prerequisites for Licensure and Professional Practice [online]. Available: http://www.asce.org/Content.aspx?id=26194 (accessed January 5, 2011).

Brierley, G., and A. Hawks. 2010. Tunnel education in the United States. Mining Engineering. 62(6):150.

EPA (U.S. Environmental Protection Agency). 2008. Wire and Cable Insulation and Jacketing: Life-Cycle Assessments for Selected Applications. EPA 744-R-08-001. U.S. Environmental Protection Agency, June [online]. Available: http://www.epa.gov/dfe/pubs/wire-cable/lfs-lca-chap1.pdf (accessed October 2, 2012).

Hapgood, F. 2004. The Underground Cutting Edge: The innovators who made digging tunnels high-tech. Invention and Technology 20(2) [online]. Available: http://www.mindfully.org/Technology/2004/Underground-Boring-Machines1sep04.htm (accessed October 2, 2012).

NAE (National Academy of Engineering). 2004. The Engineer of 2020: Visions of Engineering in the New Century. Washington, DC: The National Academies Press.

NRC (National Research Council). 1989. Geotechnology: Its Impact on Economic Growth, the Environment, and National Security. Washington, DC: National Academy Press.

PSE (Puget Sound Energy). 2012. Snoqualmie Falls Hydroelectric Project [online]. Available: http://pse.com/inyourcommunity/king/Pages/Snoqualmie-Falls.aspx (accesses July 31, 2012).

Appendixes

A

Committee and Staff Biographies

Paul H. Gilbert (NAE) retired as senior vice president, principal professional associate, and principal project manager of Parsons, Brinckerhoff, Quade, and Douglas, Inc., senior vice president of Parsons Brinckerhoff International Inc., and director of Parsons Brinckerhoff, Inc. A member of the National Academy of Engineering, his expertise is in project management of the design and construction of large complex facilities, including major subterranean constructed works. Mr. Gilbert was the project director of the Parsons Brinckerhoff/Morrison Knudsen team for design, construction management, and construction of the conventional facilities of the Department of Energy's superconducting super collider, which included 72 miles of tunnels. He has served as principal-in-charge for major engineering projects such as the Stanford Linear Accelerator, Positron-Electron Project, the Basalt Waste Isolation Project at Hanford, Washington, the Downtown Seattle Transit Subway Project, the Long Beach Naval Fuel Pier, and the Boston and San Francisco Effluent Outfall Tunnels. He is the author of Parsons Brinckerhoff's Project Management Manual and has also published various technical papers and articles. Apart from Parsons Brinckerhoff, as an appointed member of the University of California President's Council, Mr. Gilbert Chaired the Council's Project Management Oversight Panel providing project management oversight, support and mentoring over a 10 year period for large value projects executed at the Los Alamos National Laboratory, Lawrence Livermore National Laboratory and Lawrence Berkeley National Laboratory, including the National Ignition Facility at Livermore. He is also the Chair of the Associated Universities Inc. Project Management Oversight Committee for the construction of the ~$600M North American contribution to the Atacama Large Millimeter/submillimeter Radio Astronomy Array being constructed in Chile, South

America, at a site located in the Atacama Desert at 16,500 ft elevation. Mr. Gilbert is a Licensed Professional Engineer in 17 states and is a member of a variety of professional organizations, including the American Society of Civil Engineers, The Moles, and the Society of American Military Engineers. He has won multiple awards in civil engineering and construction management, including American Society of Civil Engineers fellow, its Rickey Medal, and Construction Management Award. Mr. Gilbert received his B.S. and M.S. degrees from the University of California, Berkeley in civil engineering and structural mechanics, respectively and is a Distinguished Engineering Alumnus.

Samuel T. Ariaratnam is a Professor in the Del E. Webb School of Construction in the School of Sustainable Engineering and The Built Environment (SSEBE) at Arizona State University. His teaching and research interests are in the areas of Urban Infrastructure Management & Rehabilitation, with a particular focus on trenchless engineering applications of horizontal directional drilling, trenchless pipe replacement, and underground utility asset management. Prior to joining ASU, Dr. Ariaratnam served for five years in the Department of Civil and Environmental Engineering at the University of Alberta. He has also served as a visiting assistant professor at the U.S. Air Force Academy in Colorado Springs, and, while still a graduate student, was employed at the U.S. Army Corps of Engineers Construction Research Laboratories where he performed research in military construction and strategic planning. He has published over 200 technical papers in refereed journals and conferences, has co-authored five textbooks, and is a co-holder of three patents. Dr. Ariaratnam serves as the Chairman of the International Society for Trenchless Technology and is active in a number of professional societies. He has received multiple awards including the prestigious ASCE John O. Bickel Award; the Young Civil Engineer Achievement Award from the University of Illinois, and an award of recognition from Halliburton Energy Services for contributions to underground technology. In 2006, he was named to the Phoenix Business Journal's Forty under 40 list. Recently, Trenchless Technology Magazine named him as the 2012 Trenchless Person of the Year. Dr. Ariaratnam holds a B.A.Sc. from the University of Waterloo (Canada) and an M.S. and Ph.D. in civil engineering from the University of Illinois at Urbana-Champaign.

Nancy Rutledge Connery has worked to advance civil infrastructure systems for nearly 30 years. Her career began as a transit analyst in the New York City Mayor's Office of Management and Budget at the outset of the transit system's historic reconstruction. Later, in her home state of Washington, she developed the Public Works Trust Fund, a nationally recognized program, which has provided over $2.5 billion in low interest loans for local renewal projects since 1986. She also worked as an investment banker with Seattle Northwest Securities. In 1985, she was named Executive Director of the National Council on Public Works Improvement, a joint Presidential/Congressional study commission, where she produced a series of well-regarded reports and frequent testimony on the state of the nation's infrastructure. In 2000, U.S. Senate Majority Leader appointed Connery to the Amtrak Reform Council, a financial oversight board. She has served on various National Academies' technical boards, the Executive Committee of the Institute for Civil Infrastructure Systems at New York University, and the Advisory Board of the Taubman Center for State and Local Government at Harvard Kennedy School. She has published, lectured and consulted widely throughout the world and is currently researching "next generation" infrastructure design. She holds a B.A. in political science from Pacific Lutheran University in Tacoma, Washington, and M.P.A. from John F. Kennedy School of Government, Harvard University.

Gary English is Deputy Chief of the Seattle Fire Department and Assistant Fire Marshal for the City of Seattle, Washington. He serves as the department's command staff assigned to ensure code and standard compliance with minimum fire and life safety requirements of major projects including the Sound Transit Light Rail Link combined bus/rail tunnel complex and associated stations, and Alaska Way Viaduct replacement, a stacked road tunnel to be completed by 2015. These projects include the installation of a multitude of fire and life safety systems such as point extraction ventilation, sprinkler systems, fire alarms, mass notification systems, emergency exiting, and intelligent traffic systems. In addition to such projects, Chief English is heavily involved with standard building fire and life safety systems with specialization in high-rise structures, smoke management, fire alarm, and elevator use. Chief English is a member of the International Fire Fighters Association, has served on National Fire Protection Association technical committees for road tunnels (NFPA 502), passenger rail tunnels (NFPA 130), and standpipe systems (NFPA 14), has received special training in national incident management systems at the command staff level, and has presented internationally on Underground Command and Safety.

Conrad W. Felice is the Technical Director of Tunnel Service and Vice President at HNTB Corporation and an Adjunct Professor at the University of Florida in the Department of Civil and Coastal Engineering. He has held CEO/President and Vice President level positions in multinational corporations and served in the Air Force for 27 years, retiring the rank of Lieutenant Colonel. He is a registered Professional Engineer in twelve US states, Puerto Rico and two provinces in Canada. Dr. Felice's underground and tunneling experience includes the program management of the Department of Defense underground technology development program and the technical direction for laboratory testing and analysis in support of the U.S. underground nuclear testing program at the Nevada Test Site. Commercial projects have included the physical vulnerability of underground systems to explosive and fire loads, the stability of historic road and rail tunnels, geotechnical analysis and support requirements for transit and water conveyance tunnels, seismic analysis and upgrade of underground systems, and the design and construction of large diameter pipelines underground in urban and mountainous environments. Dr. Felice has been recognized as an invited member to the National Academy of Engineering Symposium on Frontiers of Engineering and was also selected as a member of the American Society of Civil Engineers Reconnaissance Team visits to the affected earthquake damaged area of the Sichuan, China basin. He is the past Chair of the Transportation Research Board Committee on Modeling for the Design, Construction and Management of Geosystems and is serving on the Transportation Research Board Committee on Tunnels and Underground Structures as well is an active member in the International Tunneling Association working group on research. Dr. Felice earned his B.S. from Ohio University, an M.S. from the Air Force Institute of Technology and a Ph.D. in Civil Engineering from the University of Utah.

Youssef Hashash is a Professor in the department of Civil and Environmental Engineering at the University of Illinois which he joined in 1998. His research interests include deep excavations, earthquake engineering, numerical modeling, and soil-structure interaction. He is also involved in the use of visualization and virtual reality techniques in geotechnical engineering applications. Dr. Hashash worked as a staff engineer for the Parsons Brinckerhoff/Morrison Knudsen team in Dallas, Texas on the Superconducting Super Collider Project construction and was part of the Geotechnical and Underground Engineering group at Parsons Brinckerhoff in San Francisco, California. He has been involved in many tunnel and deep excavation projects around the United States and Canada. Dr. Hashash is a member of the American Society of Civil Engineers (ASCE), the Earthquake Engineering Research Institute, the American Underground Association, the International Tunneling Association, and serves on the Earth Retaining Structures Committee of the Geo-Institute of ASCE, and Performance of Structures during construction of SEI. In 2002 Dr. Hashash was named a Beckman Fellow at the Center of Advanced Studies at the University of Illinois. He is a 2001-2003 American Bridge Faculty scholar (UIUC). In 2000, Dr. Hashash is a recipient of the Presidential Early Career Award for Scientists and Engineers (PECASE) and the Walter L. Huber Civil Engineering Research Prize and Arthur Casagrande Professional Development Awards from the Geo-Institute of ASCE. He was twice a National Center for Supercomputing Application Fellow (UIUC). He received the James Crose Medal (ASCE, 1994) and Thomas Middlebrooks Awards (ASCE, 1997) for journal publications. Dr. Hashash earned his B.S., M.S., and Ph.D. degrees in Civil Engineering from the Massachusetts Institute of Technology.

Chris Hendrickson (NAE) is the Duquesne Light Company University Professor of Engineering, Co-Director of the Green Design Institute at Carnegie Mellon University, and Editor-in-chief of the American Society of Civil Engineering (ASCE) Journal of Transportation Engineering. He was elected to the National Academy of Engineering in 2011 "for leadership and contributions in transportation and green design engineering." His research, teaching, and consulting are in the general area of engineering planning and management, including design for the environment, project management, transportation systems, finance, and computer applications. Current research projects include lifecycle assessment methods (especially based on economic input/output tables such as eiolca.net), assessment of alternative construction materials, economic and environmental implications of Ecommerce, product takeback planning, and infrastructure for alternative fuels. He has co-authored three textbooks, Environmental Life Cycle Assessment of Goods and Services: An Input-Output Approach (Resources for the Future, 2005), Project Management for Construction (Prentice-Hall, 1989) and Transportation Investment and Pricing Principles (John Wiley & Sons, 1984) and two monographs, Knowledge Based Process Planning for Construction and Manufacturing (Academic Press, 1989) and Concurrent Computer Integrated Building Design (Prentice-Hall, 1994). In addition, he has published numerous articles in the professional literature. Prof. Hendrickson is a Distinguished Member of the ASCE, an Emeritus Member of the Transportation Research Board, and a Fellow of the American Association for the Advancement of Science. He has been the recipient of the 2002 ASCE Turner Lecture Award, the 2002 Fenves Systems Research Award, the 1994 Frank M. Masters Transportation Engineering Award, Outstanding Professor of the Year Award of the ASCE Pittsburgh Section (1990), the ASCE Walter L. Huber Civil Engineering Research Award (1989), the Benjamin Richard Teare Teaching Award (1987), and a Rhodes Scholarship (1973). Dr. Hendrickson earned his B.S. in General Engineering and M.S. in Civil Engineering, both from Stanford University. He earned a B.Phil. in Economics from Oxford University, and his Ph.D. in Civil Engineering from the Massachusetts Institute of Technology.

Priscilla P. Nelson is a Professor of civil engineering at the New Jersey Institute of Technology (NJIT). She served as NJIT's provost and senior vice president of academic affairs from 2005 – 2009 and then rejoined the faculty as an active professor and researcher. Prior to her tenure at NJIT, she held many positions at the National Science Foundation (NSF), concluding service there as senior advisor to the director of the National Science Foundation. Prior to her appointment to NSF, Dr. Nelson was professor of civil engineering at The University of Texas at Austin. Dr. Nelson has a national and international reputation in geological and rock engineering and the application of underground construction. She is former president of the Geo-Institute of the American Society of Civil Engineers (ASCE), a lifetime member, Fellow and first president of the American Rock Mechanics Association, a Fellow of American Association of the Advancement of Science, a Distinguished Member of ASCE, and she served on the Executive Committee of the American Geological Institute. In addition to these activities, she has many other professional affiliations including: Tau Beta Phi, the Moles, Underground Construction Association (SME), Association of Engineering Geologists, International Tunneling Association, Dispute Review

Board Foundation, and the American Society for Engineering Education. She has served as a member of and liaison to many National Research Council boards and committees. Dr. Nelson has been a part of several major construction projects, including field engineering responsibilities during construction of the Trans-Alaska Pipeline System, and serving as a consultant to the U.S. Department of Energy and the State of Texas for the Superconducting Super Collider project. She was appointed a member of the Nuclear Waste Technical Review Board by President Clinton in 1997 and again in 2000. She has authored over 125 technical publications, and received many awards, including the Kenneth Andrew Roe Award, American Association of Engineering Societies in 2008 and the Henry L. Michel Award for Industry Advancement of Research, American Society of Civil Engineers, 2011. Dr. Nelson received her Ph.D. from Cornell University in geotechnical engineering.

Raymond L. Sterling is a Professor Emeritus at Louisiana Tech University. From 1995 to 2009 he was the Contractors' Educational Trust Fund Professor of Civil Engineering and Director of the Trenchless Technology Center at Louisiana Tech University. Previously, from 1977 to 1995, he was the founding director of the Underground Space Center at the University of Minnesota. He is a Past Chairman of the International Society for Trenchless Technology and the North American Society for Trenchless Technology, a Past Chairman of the U.S. National Committee on Tunneling Technology, and a Past Animateur for the International Tunnelling Association's Working Group on Direct and Indirect Advantages of Underground Structures. In 2003, he received the Stephen D. Bechtel Pipeline Engineering Award from the American Society of Civil Engineers. He was selected as the Person of the Year by the Trenchless Technology Magazine in 2001 and as Most Valuable Professional by the Gulf Coast Trenchless Association in 1999. In 2009, he received the Gold Medal from the International Society for Trenchless Technology for outstanding contributions to the field. He has authored approximately 200 books, technical papers and reports on a wide range of topics related to underground space use, underground construction, geomechanics and trenchless technology, and for the past 10 years has served as a Senior Editor of the international journal Tunnelling and Underground Space Technology. He is registered engineer in the United States and is a Fellow of the American Society of Civil Engineers. Dr. Sterling received a B.Eng degree in civil and structural engineering from the University of Sheffield. He received his M.S. and Ph.D. degrees from the University of Minnesota.

George J. Tamaro (NAE) is a retired partner at Mueser Rutledge Consulting Engineers. His technical interests are primarily in structural and geotechnical engineering. His work involves a broad range of analytical, design, and construction problems related to deep foundations and underground structures, and he is also involved in the design and construction of containment facilities and the control of dam seepage using special barrier systems. Mr. Tamaro holds several patents in applications of slurry wall and slurry trench technology. Mr. Tamaro has an interest in the preparation and training of young engineers who will someday be consultant engineers. He is particularly concerned with the development of engineers capable of analyzing, designing, and installing safe, economically constructed facilities. He is a member of NAE, recognized for his expertise in the design and construction of slurry walls and deep foundations worldwide. Mr. Tamaro earned a B.S. degree in civil engineering from Manhattan College, and M.S. degrees in civil engineering from Lehigh University and in architectural technology from Columbia University and was awarded an honorary D. Eng. From Manhattan College.

Fulvio Tonon is Assistant Professor, Department of Civil, Architectural and Environmental Engineering at the University of Texas at Austin. He joined the faculty of the University of Texas at Austin in 2005 after spending three years as an assistant professor of geological engineering at the University of Utah, and two years as a senior tunnel engineer with Parsons Brinckerhoff. He directs the International Tunneling Consortium, which encourages academic research in response to industrial needs, after its official launch in fall 2007. He also developed an On-line Certificate in Tunneling, which aims at providing working knowledge in design or construction management of tunneling projects to UT graduate students and the industry; the program has received provisional endorsement by the International Tunneling Association (ITA). Dr. Tonon has established from scratch a rock mechanics laboratory for the characterization of intact rock and fractures as well as index tests for estimating the penetration and abrasion rate of tunnel boring machines. In 2006, Dr. Tonon won the Award for Applied Rock Mechanics from the American Rock Mechanics Association for his paper "Stresses in anisotropic rock masses: an engineering perspective building on geological knowledge". His research emphasizes rock mechanics and engineering, underground excavations and uncertainty modeling with generalized theories of probability. He has published two books on tunneling, one book on uncertainty bounds in civil engineering, 55 papers in peer-reviewed journals and 44 papers in conference proceedings. He has more than 15 years of professional experience working on projects in the Americas, Europe, and Africa. Design experience includes: cut-and-cover and bored tunnels in rock, soft ground and mixed face conditions, with or without the use of Tunnel Boring Machines; foundations and special foundations; rock and soil slope stabilizations; precast concrete and steel-concrete composite bridges; hydraulic infrastructures for dams, purification plants and rivers; renovations of ancient masonry buildings; and reinforced concrete buildings. Dr. Fulvio Tonon earned his Laurea in civil engineering from the University of Padova, Italy, and his Ph.D. in civil engineering from the University of Colorado at Boulder.

B

Open Session Meeting Agendas

MEETING 1

Committee on Underground Engineering for Sustainable Development
National Research Council of the National Academies
Meeting, June 1, 2010
AGENDA

The Keck Center of the National Academies,
500 Fifth Street, NW, Washington, D.C.

9:30 a.m. Welcome, Introductions, *Paul Gilbert, NAE*, Committee Chair

9:35 Discussion: The Definition of "Sustainability"

10:00 Features, Functions, and Characteristics of Underground Spaces: Large Occupancy Underground Environments in Europe, Asia and the U.S.
Mr. Kevin Peterson, Architect, Peterson Design

11:00	Attributed 3D Geologic Models in the UK and their Application to Planning and the Sustainable Development of Underground Space *Dr. Helen Reeves*, British Geological Survey
12:00 pm	Working Lunch in meeting room Continuation of Discussion
1:00	Sustainability Issues for Underground Space in Urban Areas *Dr. Ray Sterling*, Louisiana Technical University (by phone) (Note: Paper included in briefing materials; please be prepared to discuss)
2:00	Break
2:20	Lifecycle Costs and Benefits Issues with Urban Underground Facilities *Dr. Harvey Parker*, Harvey Parker and Associates, Inc.
3:20	Materials Issues for Underground Construction *Dr. Edward Garboczi*, National Institute for Standards and Technology
4:00 pm	Adjourn Open Session

Remainder of meeting held in closed session.

MEETING 2

Committee on Underground Engineering for Sustainable Development
National Research Council of the National Academies
July 27, 2010
AGENDA

Arnold and Mabel Beckman Center of the National Academies
100 Academy Drive, Irvine, CA

Objective: Gather information through panel discussions on issues associated with 3D planning and zoning and with operations and maintenance of underground infrastructure.

12:45 p.m. Welcome and Introductions, *Paul Gilbert* (NAE), Committee Chair

Format of panel discussions:
The panel discussions will each last approximately 2 hours. Each speaker will have approximately 15 minutes to speak followed by 5 minutes of questions and answers directed at that speaker. Following the last speaker, the table will be open for roundtable discussion. All present are invited to participate in discussion.

12:50 p.m. Panel 1—Three-Dimensional Urban Planning
Moderator: Priscilla Nelson, PhD, Committee Member

Panelists:
Richard Little, Director, Keston Institute for Public Finance and Infrastructure Policy, University of Southern California
Gordon Feller, Director, Internet Business Solutions Group, Cisco Systems
Benedict Schwegler, Jr., PhD, Vice President and Chief Scientist, Walt Disney Imagineering Research and Development

2:45 Break

3:00 Panel 2—Operating and Managing Underground Space
Moderator: Samuel Ariaratnam, PhD, Committee Member

Panelists:
F. G. Wyman Jones, Manager and Supervising Engineer, Los Angeles County Metropolitan Transit Authority
Henry A. Russell, Vice President, Parsons Brinkerhoff, Inc., Boston
Razmik Manoukian, Los Angeles Department of Water and Power

5:00 Reception for committee and guests on the Dining Terrace

5:30 Working dinner for committee and guests in the Executive Dining Room
Discussion over dinner regarding the day's panel sessions

7:00 p.m. Adjourn

Remainder of meting held in closed session.

C

Interdisciplinary Underground Engineering Practice

This appendix focuses on an integrated systems approach to the practices of urban planning, underground development, and maintenance and discusses the multidisciplinary effort required to manage underground infrastructure in an integrated way throughout infrastructure lifecycle. Contracting practices are also discussed.

THE INTERDISCIPLINARY UNDERGROUND ENGINEERING TEAM

Underground engineering is a multidisciplinary endeavor given the challenges associated with creating healthy, safe, productive, and pleasant space in complex geologic environments and in inhospitable and dangerous conditions. A range of considerations—from social, political, and economic, to physical condition of the ground, to environmental preservation, to those related to human factors—play into decisions to be made by the necessarily interdisciplinary teams that plan, construct, operate and maintain. Some specialties tapped during planning and construction are described in this appendix to demonstrate the complexity of engineering underground infrastructure with sustainable development in mind. The sections are broken into distinct phases or specialty areas, but successful project completion depends on an integrated approach and constant interaction among team members.

Early Planning for Sustainability

During early underground infrastructure planning for a representative project, urban and regional planners consider opportunities to optimize the surface environment through underground use. Specialized site and facility planning specialists consider the interrelationship between the surface and underground in detail. Architects, architectural engineers, and underground civil engineering specialists develop and illustrate workable design solutions for specific underground spaces that reflect geologic and groundwater conditions and any hazards present at and near the site of the proposed facility. Geotechnical engineers provide site definition investigations and studies that establish the basis for these designs (see below). To appreciate what the finished facility may look like and how well it will serve the intended purposes, interior designers will consider how to transform the underground space by planning surfaces, lighting, colors, finishes, textures, signs, and subliminal indicators that contribute to a sense of comfort and safety in the underground and its public access points.

Cost Estimating, Schedule Management, and Interface Management

A component of project "success" is completion within cost budgets and time schedules. Cost and schedule management specialists develop workable schedules and correlated cost estimates for the underground work that reflect reported site conditions and constraints, equipment selection to perform the work, and any mitigations designated by the various government offices with jurisdiction. As the project advances to construction, the leads on the interdisciplinary team, define major discrete work elements and implement control systems that assist in the management of project scope, costs, and schedules beginning with a project work breakdown structure (WBS). They coordinate cost estimates, time schedules, and data that form baseline working budgets for each task and work package. A critical path schedule is developed to aid work management. The assembled information becomes a regularly updated project control system that reflects progress and identifies occurrences of indications of problems or delays as early as possible. Underground project cost estimating specialists conduct risk and contingency analyses, analyze processes and scheduling, and recommend the most cost effective equipment for specific jobs. They rely on the work of underground systems engineering specialists who develop a project risk register[1] used in project planning, risk assessment, and risk mitigation.

Systems engineering specialists, besides developing risk registers, employ interface management techniques that integrate project management areas and technical disciplines. As a project develops, construction contract packages are defined for different parts of the project, and interfaces between the separate

[1]A risk register is a tool created shortly after a project concept is defined, and is used to manage risk in underground construction and operation throughout project development. It helps to identify risks and their impacts, from which mitigating and contingency actions can be determined. Many risks listed in the register (e.g., discipline-specific uncertainties such as the availability of specific electronic devices) are reduced as work progresses and mitigation activities are performed.

contracting packages coming from different contractors must be identified and actively managed. Other systems engineering assignments may include configuration management (assuring system function and performance are consistent with design) and change control management. A small, well-lead group of experienced procurement practitioners will prepare and proof contract documents and oversee procurement processes for the several underground contracts to be bid and built.

Site Characterization and Environmental Protection

Engineers with training in many disciplines are necessary to characterize the underground, re-engineer existing infrastructure, and create the environment necessary to support the proposed underground infrastructure. The expertise of civil engineers trained in the design, construction, and maintenance of public works is augmented with expertise of those with graduate training in areas such as geotechnical and geological engineering, rock and soil mechanics, and geophysics.

Geotechnical engineers classify the engineering behavior of earth materials through field and laboratory testing. Geological engineers interpret how the geology and geologic origins of an area might influence planning, mining, construction, and operation of the infrastructure. Understanding the soil and rock mechanical properties at and near a job site is necessary for smart and safe design and construction of proposed underground structures. Rock mechanics engineers describe expected in situ strengths, stresses, strains, elasticity, and other rock mass properties where work will occur. Soil mechanics engineers similarly describe the nature and behavior of the less consolidated materials—soils—in the area. In the early design and development of the project, these various specialists prepare a geotechnical baseline report (GBR),[2] take part in specifying and selecting tunnel boring machines (TBMs), and, as designs develop, work with the design and specification drafting team. Geophysical engineers noninvasively measure physical properties of an area using equipment and analytical techniques to infer geology, geologic structure, groundwater conditions, to identify geologic anomalies where they exist, and the presence of manmade artifacts or potentially dangerous gases. This aids production of three-dimensional models of soil and rock characteristics. Gaseous ground mitigation specialists will design measures to mitigate and control each type of gas.

Groundwater Protection and Control

Environmental engineers and planners identify potential impacts to the environment associated with underground infrastructure construction and work with

[2] A GBR is used to define the baseline conditions on which contractors will base their bids and select their means, methods, and equipment (FHWA, 2011a).

design engineers to identify appropriate construction plan, operation and mitigation measures. An environmental impact report or statement is completed before the design of any project begins in earnest, and mitigation action plans—as needed—will be part of submitted planned contract bid packages. Project delivery that includes mitigation measures is a means to assure environmental protection.

Groundwater hydrology engineers address challenges associated with protecting groundwater as a resource, and with engineering in consideration of groundwater as a component of the soil and rock. The control of groundwater, commonly encountered during underground construction, is a complicated and potentially costly physical challenge. It is important that groundwater conditions at infrastructure works shafts and surface penetrations are mapped and analyzed, and that groundwater preservation and management solutions are developed and coordinated with all team members. Groundwater conditions can influence a series of fundamental decisions related to construction and operation, including choice of ground excavation methods, support systems, and of water barriers that may be required to protect groundwater resources and maintain a "dry" underground facility.

Facility Construction

Once construction begins, structural engineers lead units charged with the design of specific project sections or elements (normally what becomes a construction contract package). These engineers know how appropriate connections between the different structural components are made, and may have to incorporate seismic design and waterproofing concepts into underground structures to make structures safe for given circumstances. Seismic engineers evaluate for and prepare site-specific seismic designs criteria to assure compliance with existing codes and safety. Geotechnical engineers, TBM specialists, and mining engineers may be enlisted as part of the team to allow safe excavation of materials from within the earth.

Building underground facilities, as with many aboveground facilities, requires the expertise of underground works construction engineers who understand problems associated with crowded work sites and difficult logistical considerations. Work sequences can become disordered by small events that result in delays and added costs. A seasoned construction engineer performs full constructability reviews to anticipate such problems as the designs develop.

Mechanical, Electrical, and Communication Systems

Trained mechanical engineers design, construct, and operate, multiple mechanical systems needed in underground projects, including water management (e.g., sumps and piping, valves, pumps and motors, and controls), and heating, air conditioning, and ventilation systems. Mechanical equipment needs to be selected consistent with existing codes and standards, with energy efficiency in mind, and in consideration of the underground environment and sustainable use. Underground safety is enhanced through application of computational fluid dynamics techniques to determine how fluid and gasses flow in and around underground spaces. Noise mitigation and the safety and comfort of occupants of the underground are also of concern to mechanical engineers. Electrical and communications systems engineers design power transmission, distribution and communications systems, and the surveillance and security sensors, systems, and control rooms from which these systems are operated. Electrical systems are essential to the work of many other engineers on a given project, and to safe infrastructure operation.

Fire Protection and Life Safety

Assuring the safety of people underground requires accommodating the physical needs of survival. Ventilation experts design systems to provide fresh air and remove excess carbon dioxide and other gases that build up in enclosed spaces. Fire safety specialists understand how underground fire may start, spread, be contained, and extinguished, and they understand how smoke, heat, or hazardous materials may travel through underground passageways and pose threats. They also understand how many emergency evacuation routes are needed and where they can be placed, and under what circumstances shelter-in-place facilities might be appropriate. Each underground project is under the jurisdiction of fire marshals who guide infrastructure design to satisfy fire and life safety requirements, however safety and sustainability are often dependent, especially in the underground, on moving beyond regulatory compliance.

Although not generally considered a primary purpose in the United States, underground infrastructure may be called upon to shelter people against natural or manmade hazards. Shelter design engineers or weapons effects experts may be asked to integrate the requirements of such use into facility design.

UNDERGROUND CONSTRUCTION CONTRACTING PRACTICES

Underground construction contracting, as with all types of construction contracting, ranges from small and simple to large and complex. Project contracting differs depending on the scope, work site location, sources of funding, if the project will be built by union employees or "open shop", and applicable laws, ordinances, and regulations. Contract provisions unique to underground construction practices, however, have been developed according to site geology and groundwater conditions, site uncertainties and risks, special project insurance provisions, and payment terms corresponding to risks shared by the owner and contractor. Such provisions have been formalized with support of the underground construction industry and successive blue ribbon committees comprised of experts from federal agencies, the engineering and construction industry, academe, and the legal services and insurance industries. The National Research Council's U.S. National Committee on Tunneling Technology (USNCTT) produced reports from 1974 through 1995 covering contracting and technical issues intended to improve the performance of underground construction (NRC, 1974; 1977; 1978; 1984; 1989; and 1995).

Summaries of Key USNCTT Issues

The body of work from the USNCTT improved contracting and management practices for underground construction in the United States. Their reports are still important resources today, remaining relevant in the face of derivative laws of the National Environmental Policy Act of 1969[3] that cover water, air, noise, endangered species, remediation, as well as the introduction of much new technology. The value of the work of the USNCTT in improving contracting practices and management for underground facilities is illustrated in the following summaries of issues addressed in their reports. The American Society of Civil Engineers (ASCE 1997, 2007), the Underground Construction Association[4] and the International Tunneling Association (e.g., ITA, 1988) more recently have addressed improvements to construction contracting for underground works. Below are descriptions of key reports.

- *Better Contracting for Underground Construction* (NRC, 1974) provides recommendations pertaining to:

    - Full disclosure to all bidders of all subsurface information, professional interpretations, and design considerations through a special report of geotechnical site conditions and facility designs, with careful distinction given to what was factual data and what were interpretations or opinions. The report spoke out against the use of disclaimers by owners pertaining to underground geotechnical data provided.
    - Provisions for a changed-conditions clause within contracts to include differing-site-conditions thus identifying for the owner where it might assume the risk concerning unknowns in subsurface physical conditions.
    - Provisions for contingencies for special groundwater problems.
    - Encouragement of the use of cost-reimbursement contracts for major underground construction with particular terms and conditions.
    - Provisions for bidder pre-qualifications.
    - Provisions for use of value engineering.

[3] See www.epa.gov/lawsregs/laws/nepa.html.
[4] See http://uca.smenet.org/.

- Provision for the use of bid pricing to provide for timely payment for up-front mobilization and other expenses for the contractor, and alternative bidding for private work but not for public works (contrary to then-existing public laws).
- Provisions covering cost escalations during the contract period.
- Provisions to substantially reduce the time and cost of submitting, negotiating and obtaining payment of contract price adjustments for approved changes.
- Disclosure of the engineer's estimate for the cost of the construction as-bid with the opening of bids; and, disclosure with the invitation for bids, providing notice given a limit on the total funding available for the contract.
- Provisions for wrap-up project insurance.
- Provisions for use of the arbitration process, a process developed by the American Arbitration Association (AAA) to settle disputes under the contract, short of litigation.

These recommendations left open the use of disclaimers by an owner when providing site geotechnical data and information for prospective underground construction bidders and construction managers. This proved to be a significant opportunity for disputes and litigation centered on this issue, which the committee was soon to address.

- *Recommended Procedures for Settlement of Underground Construction Disputes* (NRC, 1977) detailed the processes for employing the AAA procedures that included in contracts:

  - Provisions to employ AAA's underground construction disputes settlement rules to settle disputes equitably and cost effectively through the use of both mediation and arbitration.
  - Provisions for either party, if in their interest, to call for the matter in dispute to go to arbitration, with the knowledge that the arbitration report would be entered into evidence should the matter ultimately go to litigation.

The use of the mitigation or arbitration process brought into focus the need for experienced underground construction experts who could objectively hear the arguments for and against a disputed issue and find a suitable resolution based on the contract, technical considerations, the law and precedence and also be versed in the process. If mitigation was employed, a single party acceptable to both could handle the process and matter. However if the matter was taken to arbitration, the process typically would entail a panel of three persons. This process initially worked, but was costly, time consuming, and the number of people qualified for the processes was limited.

This led to the advent and use of construction-industry organized and populated, Dispute Review Boards (DRB), (called Disputes Resolution Boards in some states) based on the form of the arbitration process, but entirely staffed with senior and semi-retired experienced underground construction personnel, first used in 1975. Their proceedings were found to take less time, cost less, to be more amicable, and more equitable in their outcome. Thereafter the Disputes Resolution Board Foundation was formed to provide a central source of information and optional form-contract terms for use across the United States and throughout the world.

- *Better Management of Major Underground Construction Projects* (NRC, 1978) observed that underground projects are among the most complicated because they typically take place in urban areas, geotechnical considerations assume greater importance than in other types of construction, and underground work requires special equipment, techniques and skills to perform. The success of underground projects is therefore particularly sensitive to the management practice employed. The report provides recommendations to address and avoid management problems of projects through the use of examples. The report identifies that the most important cause of management problems is delay in taking decisive action. A list of goals and objectives to improve the management of underground projects is provided and detailed.
- *Geotechnical Site Investigations for Underground Projects* (NRC, 1984) which provided an entirely new strategy for improving contractor/owner relationships and outcomes related to underground projects. This was a strategy for fair risk sharing, particularly of those associated with unknown ground conditions. Complete disclosure of all factual geotechnical information to all bidders, and the preparation of a special report that documents the designer's reasoning and interpretations that resulted in the selection of construction methods, lining types, anticipated ground behavior and other information was recommended.

Central to this new contracting strategy was an appendix to the report entitled *Geotechnical Design Report (*NRC, 1984). With the use and purpose remaining the same, practitioners in the industry found it more convenient to rename it "Geotechnical Design Summary Report" (GDSR). In 1997, the ASCE through its Underground Technology Research Council, Technical Committee of Contracting Practices expanded and added clarity and form of the GDSR document, revising the title to *Geotechnical Baseline Report* (GBR) (ASCE, 1997). ASCE and others have continued to improve the GBR in application (e.g., Smith, R.E., 2001; van Staveren and Knoeff, 2004; ASCE, 2007; and Rozek and Loganathan, 2008). There is a continuing need to update the content of the GBR and to anticipate the geotechnical and design requirements for use of new technology in ground improvement tools and materials, new mining tools and capabilities, etc. coming into the construction processes. Ongoing research is indicated.

In 1989 the USNCTT addressed the anticipated construction of the Superconducting Super Collider (SCC; 72 miles of tunnels, 42 shafts, 4 large underground rooms, etc.). Recommended contracting practices for this project, addressed effective use of rigorous risk recognition, risk assessments and minimizing, and allocation of risks on the basis of what party can best manage the risk or consequences.

The SSC project team was to identify then mitigate, minimize, control, or eliminate project risks to the performance of the underground work. These recommendations were applied by the project team at every step with the outcome that with over $400 million in construction completed or in work under firm fixed price construction contracts, the project was $100 million under the project's target cost for that scope when the government terminated the project in 1993 (P. Gilbert, personal communication).

- *Safety in the Underground Construction and Operation of the Exploratory Studies Facility at Yucca Mountain* (NRC, 1995) issued its last report on the topic of safety. It is interesting that the always significant issue of underground work safety was addressed in this last USNCTT publication.

The beneficial contributions made to U.S. underground construction practices cannot be overstated. Revisiting the need for a body that could provide unifying guidance related to the full array of underground infrastructure issues for the lifecycle of said infrastructure may be appropriate, especially with respect to how underground engineering practice may ultimately contribute to urban sustainability.

REFERENCES

ASCE (American Society of Civil Engineers). 1997. *Geotechnical Baseline Reports for Underground Construction: Guidelines and Practices*. Reston, VA: American Society of Civil Engineers.

ASCE. 2007. *Geotechnical Baseline Reports for Construction – Suggested Guidelines*, R.J. Essex, ed. Reston, VA: American Society of Civil Engineers.

FHWA (U.S. Federal Highway Administration). 2011. Chapter 4 – Geotechnical Reports. In *Technical Manual for Design and Construction of Road Tunnels – Civil Elements*. U.S. Department of Transportation. Available at http://www.fhwa.dot.gov/bridge/tunnel/pubs/nhi09010/04.cfm.

ITA (International Tunnelling Association).1988. ITA ecommendations on contractual sharing of risks. Tunn. Undergr. Sp. Technol. 3(2):103-140.

NRC (National Research Council). 1974. *Better Contracting for Underground Construction*. Washington, DC: National Academy of Sciences.

NRC. 1977. *Recommended Procedures for Settlement of Underground Construction Disputes*. Washington, DC: National Academy of Sciences.

NRC. 1978. B*etter Management of Major Underground Construction Projects*. Washington, DC: National Academy of Sciences.

NRC. 1984. *Geotechnical Site Investigations for Underground Projects*, Volume 1 and 2. Washington, DC: National Academy Press.

NRC. 1989. *Micro and Small Diameter Tunneling*. Washington, DC: National Academy Press.

NRC. 1995. *Safety in the Underground Construction and Operation of the Exploratory Studies Facility at Yucca Mountain.* Washington, DC: National Academy Press.

Rozek, J., and N. Loganathan. Geotechnical Baseline Report as an Underground Risk Management Tool. Parsons Brinkerhoff, Sydney Australia [online]. Available: http://www.ctta.org/FileUpload/ita/2009/papers/O-01/O-01-05.pdf (accessed October 2, 2012).

Smith, E.R. 2001. Geotechnical Baseline Reports: State of the Practice. Proceedings 36 Annual Symposium on Engineering Geology and Geotechnical Engineering, Luke, Jacobson & Werle (ads) Univerity of Nevada, Las Vegas, March 28-30.

van Staveren M.T. and J. G. Knoeff. 2004. The Geotechnical Baseline Report as Risk Allocation Tool. In *Engineering Geology for Infrastructure Planning in Europe.* R. Hack, R. Azzam, R. Charlier (Eds.). pp. 777–785. New York: Springer.

Abrahamson, D., & Bakker, A. (2016). Making sense of movement in embodied design for mathematics learning. In N. Newcombe & S. Weisberg (Eds.), Embodied cognition and STEM learning. *Cognitive Research: Principles and Implications, 1*(1), 1–13. https://doi.org/10.1186/s41235-016-0034-3

Abrahamson, D., Tancredi, S., Chen, R. S. Y., Flood, V. J., & Dutton, E. (2023). Embodied design of digital resources for mathematics education: Theory, methodology, and framework of a pedagogical research program. In B. Pepin, G. Gueude, & J. Choppin (Eds.), *Handbook of digital resources in mathematics education* (pp. 217–250). Springer. https://doi.org/10.1007/978-3-030-95060-6_8-1

Abrahamson, D., Trninic, D., Gutiérrez, J. F., Huth, J., & Lee, R. G. (2011). Hooks and shifts: A dialectical study of mediated discovery. *Technology, Knowledge, and Learning, 16*(1), 55–85. https://doi.org/10.1007/s10758-011-9177-y

Abrahamson, D., & Wilensky, U. (2005, June). *Piaget? Vygotsky? I'm game! Agent-based modeling for psychology research*. Paper presented at the Jean Piaget Society, Vancouver, Canada. https://www.netlogoweb.org/launch#https://www.netlogoweb.org/assets/modelslib/Sample%20Models/Psychology/Piaget-Vygotsky%20Game.nlogo

Abrahamson, D., Worsley, M., Pardos, Z. A., & Ou, L. (Eds.). (2022). Learning analytics of embodied design: Enhancing synergy. *International Journal of Child-Computer Interaction, 32*, 100409. https://doi.org/10.1016/j.ijcci.2021.100409

Allen, M., & Friston, K. J. (2018). From cognitivism to autopoiesis: Towards a computational framework for the embodied mind. *Synthese, 195*(6), 2459–2482. https://doi.org/10.1007/s11229-016-1288-5

Balasubramanian, S., Melendez-Calderon, A., Roby-Brami, A., & Burdet, E. (2015). On the analysis of movement smoothness. *Journal of NeuroEngineering and Rehabilitation, 12*(1), 112. https://doi.org/10.1186/s12984-015-0090-9

Bamberger, J. (1999). Action knowledge and symbolic knowledge: The computer as mediator. In D. Schön, B. Sanyal, & W. Mitchell (Eds.), *High technology and low-income communities* (pp. 235–262). MIT Press.

Clark, A., & Chalmers, D. J. (1998). The extended mind. *Analysis, 58*(1), 7–19.

Collins, A., & Ferguson, W. (1993). Epistemic forms and epistemic games: Structures and strategies to guide inquiry. *Educational Psychologist, 28*(1), 25–42. https://doi.org/10.1207/s15326985ep2801_3

diSessa, A. A. (2018). Computational literacy and "the big picture" concerning computers in mathematics education. *Mathematical Thinking and Learning, 20*(1), 3–31. https://doi.org/10.1080/10986065.2018.1403544

Donald, M. (2010). The exographic revolution: Neuropsychological sequelae. In L. Malafouris & C. Renfrew (Eds.), *The cognitive life of things: Recasting the boundaries of the mind* (pp. 71–79). McDonald Institute Monographs.

Dreyfus, H. L. (2002). Intelligence without representation: Merleau-Ponty's critique of mental representation. *Phenomenology and the Cognitive Sciences, 1*(4), 367–383. https://doi.org/10.1023/A:1021351606209

Farrera, A., & Ramos-Fernández, G. (2022). Collective rhythm as an emergent property during human social coordination. *Frontiers in Psychology, 12*. https://doi.org/10.3389/fpsyg.2021.772262

Gigerenzer, G., & Brighton, H. (2009). Homo heuristicus: Why biased minds make better inferences. *Topics in Cognitive Science, 1*(1), 107–144. https://doi.org/10.1111/j.1756-8765.2008.01006.x

Goodwin, C. (1994). Professional vision. *American Anthropologist, 96*(3), 603–633. https://doi.org/10.1525/aa.1994.96.3.02a00100

Heidegger, M. (1962). *Being and time* (J. Macquarrie & E. Robinson, Trans.). Harper & Row. (Original work published 1927)

Heider, F., & Simmel, M. (1944). An experimental study in apparent behavior. *American Journal of Psychology, 57*, 243–259. https://doi.org/10.2307/1416950

Horsey, R. (2002). The art of chicken sexing. *UCL Working Papers in Linguistics, 14*, 107–117.

Hutto, D. D., Kirchhoff, M. D., & Abrahamson, D. (2015). The enactive roots of STEM: Rethinking educational design in mathematics. In P. Chandler & A. Tricot (Eds.), Human movement, physical and mental health, and learning. *Educational Psychology Review, 27*(3), 371–389. https://doi.org/10.1007/s10648-015-9326-2

Järvelä, S., Nguyen, A., & Hadwin, A. (2023). Human and artificial intelligence collaboration for socially shared regulation in learning. *British Journal of Educational Technology, 54*(5), 1057–1076. https://doi.org/10.1111/bjet.13325

Kirsh, D. (2010). Thinking with external representations. *AI & Society, 25*, 441–454. https://doi.org/10.1007/s00146-010-0272-8

Koschmann, T., Kuuti, K., & Hickman, L. (1998). The concept of breakdown in Heidegger, Leont'ev, and Dewey and its implications for education. *Mind, Culture, and Activity, 5*(1), 25–41. https://doi.org/10.1207/s15327884mca0501_3

Kugler, P. N., & Turvey, M. T. (1987). *Information, natural law and the self-assembly of rhythmic movement*. Erlbaum.

Lehr, S. A., Caliskan, A., Liyanage, S., & Banaji, M. R. (2024). ChatGPT as research scientist: Probing GPT's capabilities as a research librarian, research ethicist, data generator, and data predictor. *Proceedings of the National Academy of Sciences, 121*(35), e2404328121. https://doi.org/doi:10.1073/pnas.2404328121

Marwan, N., Romano, M. C., Thiel, M., & Kurths, J. (2007). Recurrence plots for the analysis of complex systems. *Physics Reports, 438*, 237–239. https://doi.org/10.1016/J.PHYSREP.2006.11.001

Menary, R., & Gillett, A. (2022). The tools of enculturation. In W. D. Gray, F. Osiurak, & R. Heersmink (Eds.), Tasks, tools and techniques. *Topics in Cognitive Science, 14*, 363–387. https://doi.org/10.1111/tops.12604

Mikolov, T., Sutskever, I., Chen, K., Corrado, G. S., & Dean, J. (2013). Distributed representations of words and phrases and their compositionality. *Advances in Neural Information Processing Systems, 26*, 1–9. https://doi.org/10.48550/arXiv.1310.4546

Mondada, L. (2014). The local constitution of multimodal resources for social interaction. *Journal of Pragmatics, 65*, 137–156. https://doi.org/10.1016/j.pragma.2014.04.004

Norman, D. A. (1991). Cognitive artifacts. In J. M. Carroll (Ed.), *Designing interaction: Psychology at the human-computer interface* (pp. 17–38). Cambridge University Press.

Overmann, K. A. (2023). *The materiality of numbers: Emergence and elaboration from prehistory to present.* Cambridge University Press.

Pardos, Z. A., Rosenbaum, L. F., & Abrahamson, D. (2021). Characterizing learner behavior from touchscreen data. *International Journal of Child-Computer Interaction, 31*, 100357. https://doi.org/10.1016/j.ijcci.2021.100357

Papert, S. (1980). *Mindstorms: Children, computers, and powerful ideas.* Basic Books.

Polanyi, M. (1958). *Personal knowledge: Towards a post-critical philosophy.* University of Chicago Press.

Potega vel Żabik, K., Abrahamson, D., & Iłowiecka-Tańska, I. (2024). It takes two to OЯTHO: A tabletop action-based embodied design for the Cartesian system. *Digital Experiences in Mathematics Education, 10*(2), 189–201. https://doi.org/10.1007/s40751-024-00139-8

Proksch, S., Reeves, M., Gee, K., Transtrum, M., Kello, C., & Balasubramaniam, R. (2023). Recurrence quantification analysis of crowd sound dynamics. *Cognitive Science, 47*(10), e13363. https://doi.org/10.1111/cogs.13363

Radford, L. (2013). Three key concepts of the theory of objectification: Knowledge, knowing, and learning. *REDIMAT—Journal of Research in Mathematics Education, 2*(1), 7–44.

Rosenbaum, L. F., Kaur, J., & Abrahamson, D. (2020). Shaping perception: Designing for participatory facilitation of collaborative geometry. *Digital Experiences in Mathematics Education, 6*(2), 213–232. https://doi.org/10.1007/s40751-020-00068-2

Ryle, G. (1945). Knowing how and knowing that: The presidential address. *Proceedings of the Aristotelian Society, 46*, 1–16. https://doi.org/10.1093/aristotelian/46.1.1

Schwartz, J. L. (2007). Models, simulations, and exploratory environments: A tentative taxonomy. In R. A. Lesh, E. Hamilton, & J. J. Kaput (Eds.), *Foundations for the future in mathematics education* (pp. 161–172). Lawrence Erlbaum Associates.

Sfard, A. (2002). The interplay of intimations and implementations: Generating new discourse with new symbolic tools. *Journal of the Learning Sciences, 11*(2&3), 319–357. https://doi.org/10.1080/10508406.2002.9672142

Sinclair, N. (2024). Knowing as remembering: Methodological experiments in embodied experiences of number. *Digital Experiences in Mathematics Education, 10*(1), 29–46. https://doi.org/10.1007/s40751-023-00132-7

Tenenbaum, J., Griffiths, T., & Kemp, C. (2006). Theory-based Bayesian models of inductive learning and reasoning. *Trends in Cognitive Sciences, 10*(7), 309–318. https://doi.org/10.1016/j.tics.2006.05.009

Vesper, C., Abramova, E., Bütepage, J., Ciardo, F., Crossey, B., Effenberg, A., Hristova, D., Karlinsky, A., McEllin, L., Nijssen, S. R. R., Schmitz, L., & Wahn, B. (2017). Joint action: Mental representations, shared information and general mechanisms for coordinating with others. *Frontiers in Psychology, 7*, 2039. https://doi.org/10.3389/fpsyg.2016.02039

Wilensky, U. (1999). NetLogo. Center for Connected Learning and Computer-Based Modeling, Northwestern University, Evanston. http://ccl.northwestern.edu/netlogo

Wilensky, U., & Papert, S. (2010). Restructurations: Reformulations of knowledge disciplines through new representational forms. *Constructionism, 17*, 1–15. https://ccl.northwestern.edu/2010/wilensky_restructurations_Constructionism%202010-latest.pdf

Wolfram, S. (2002). *A new kind of science*. Wolfram Media, Inc.

V Embodied Intelligence-Unifying Principles

Part V focuses on some of the unifying principles of embodied intelligence (EI). This section discusses the transition toward embodied artificial intelligence (EAI), where intelligence is seen to arise from the physical system, its materials, and its design, rather than being solely dependent on algorithms. This shift is presented as a promising avenue to address the limitations of traditional AI, particularly in handling novel tasks and effectively utilizing interactions with the environment. Furthermore, a commentary on "diverse intelligences" broadens the perspective beyond human-centered views, suggesting that intelligence is not confined to human-like minds but can manifest in various embodied forms. It is argued that concepts such as "life," "machine," "mind," "sentience," and "robot" are more accurately understood as relationships between systems rather than fixed categories, with significant implications for the practical application and ethical considerations of the resulting interactions. This part concludes with the current trajectory and future directions within the field, while drawing on ideas discussed in the volume's beginning. These ideas include the similarities between human embodied cognition (HEC) and EAI, advocating for a reevaluation of EI that encompasses embodiment across living, hybrid, and artificial entities, moving beyond purely mind-centric perspectives.

Common Themes

Challenges remain in the implementation for EI and include the need to consider the algorithmic capabilities and the physical "embodiment" of the system. Specifically, such challenges include (chapter 17):

1. **The role of representation:** Are representations always embodied, or can they also originate from internal processes (Clark, 2013, 2015; chapter 3)?

What is the nature of EAI? Is it really "strong" embodiment (chapter 8)? One potential solution could be an architecture or configuration of neuronal pathways under a (Markovian) generative model (for discrete state spaces) that can realize both representational and nonrepresentational (dynamicist) processes (chapter 3).

2. **Concepts and conceptualization:** Large language models (LLMs) are based on a particular type of neural network that implements a statistical predictive procedure using a mechanism originally designed to improve the performance of automatic translators and that does not seem to represent the concept of meaning. For example, connections in the network might not be between actual concepts, but only between concept *labels*. Strong AI would require the creation and manipulation of concepts, not concept labels (chapter 7).
3. **Emotions and agency:** The ability to regulate body dynamics is paramount, with the form and structure designed to encourage the emergence of complex behaviors and properties. Replicating the full complexity of human emotions in AI remains a challenge. It is far more controversial and still might be regarded as a mistake in the mainstream of AI research to attribute genuine agency, awareness, or consciousness to such systems (Seth, 2025). This might be because the dominant generative AI systems operate in a "feedforward" way, conditioning their responses on user input but lacking intrinsic goals (chapter 5).
4. **Artificial neural network (ANN) potential:** ANNs still lag behind biological neural networks in complexity, adaptability, and emergent properties like consciousness and creativity (Adadi & Berrada, 2018). Research is striving to bridge this gap through innovations like neuromorphic computing (Mead, 1989), brain-computer interfaces (Nicolelis, 2001), biologically inspired AI design (Hassabis et al., 2017), and explainable AI (Adadi & Berrada, 2018; Saeed & Omlin, 2023). Neuromorphic computing, utilizing spiking neural networks, might offer a promising solution by replicating the brain's biological functions. However, accuracy, adaptability, efficiency, reliability, and security challenges persist (chapter 16).

Improvements for EAI can include biological hybrids. Xenobots are "robots" (Blackiston et al., 2021) that can be formed from living cells, with some ability for these "bots" to contribute to the creation of others. This is accompanied by an increase in the development of biohybrid systems, intersecting the key advances of biological and artificial systems (chapter 16). "Living robots"—including those with self-replicating systems, self-healing, or self-sensing—would represent a truly embodied approach to the design and creation of intelligence systems (Terryn et al., 2021).

New trends include the use of soft robotics as a tool to understand intelligence, which is a growing field. This is seen in the use of soft wearable devices for people undergoing MRI studies (Kikkert et al., 2021; chapter 16).

Conclusion

Ultimately, pursuing an expanded, transdisciplinary EI view envisions a future of diverse, interacting intelligences, requiring continued commitment to a shared knowledge base. This volume is a vital initial step toward a unified comprehension of embodiment's fundamental role in intelligence.

References

Adadi, A., & Berrada, M. (2018). Peeking inside the black-box: a survey on explainable artificial intelligence (XAI). *IEEE Access, 6,* 52138–52160. https://doi.org/10.1109/ACCESS.2018.2870052

Blackiston, D., Lederer, E., Kriegman, S., Garnier, S., Bongard, J., & Levin, M. (2021). A cellular platform for the development of synthetic living machines. *Science Robotics, 6*(52), eabf1571. https://doi.org/10.1126/scirobotics.abf1571

Clark, A. (2013). Whatever next? Predictive brains, situated agents, and the future of cognitive science. *Behavioral and Brain Sciences, 36*(3), 181–204. https://doi:10.1017/S0140525X12000477

Clark, A. (2015). *Surfing uncertainty: Prediction, action, and the embodied mind.* Oxford University Press.

Hassabis, D., Kumaran, D., Summerfield, C., & Botvinick, M. (2017). Neuroscience-inspired artificial intelligence. *Neuron, 95*(2), 245–258. http://dx.doi.org/10.1016/j.neuron.2017.06.011

Kikkert, S., Sonar, H. A., Freund, P., Paik, J., & Wenderoth, N. (2021). Hand and face somatotopy shown using MRI-safe vibrotactile stimulation with a novel soft pneumatic actuator (SPA)-skin interface. *bioRxiv*. https://doi.org/10.1101/2021.04.14.439873

Mead, C. (1989). Analog VLSI and neutral systems. *NASA STI/Recon Technical Report A, 90,* 16574.

Nicolelis, M. A. (2001). Actions from thoughts. *Nature, 409*(6818), 403–407. https://doi.org/10.1038/35053191

Saeed, W., & Omlin, C. (2023). Explainable AI (XAI): A systematic meta-survey of current challenges and future opportunities. *Knowledge-Based Systems, 263*, 110273.

Seth, A. K. (2025). Conscious artificial intelligence and biological naturalism. *Behavioral and Brain Sciences*, 1–42. https://doi.org/10.1017/S0140525X25000032

Terryn, S., Langenbach, J., Roels, E., Brancart, J., Bakkali-Hassani, C., Poutrel, Q.-A., Georgopoulou, A., Thuruthel, T. G., Safaei, A., Ferrentino, P., Sebastian, T., Norvez, S., Iida, F., Bosman, A.W., Tournilhac, F., Clemens, F., Van Assche, G., & Vanderborght, B. (2021). A review on self-healing polymers for soft robotics. *Materials Today, 47*, 187–205. https://doi.org/10.1016/j.mattod.2021.01.009

16 Embodied Artificial Intelligence: Enabling the Next Intelligence Revolution

Josie Hughes, Arsen Abdulali, Ryman Hashem, and Fumiya Iida

Introduction

The concept of computation and intelligence has historically seen significant evolution, which continues to this day (Lungarella et al., 2007; Martínez-Plumed, et al., 2018). It is shaped by our social awareness, needs, and challenges and developments in both our understanding of intelligence, its origins, and technologies that show intelligent behaviors or interactions. At a time when we need solutions to global health, climate, and societal challenges, there is a pressing need to advance the capabilities of intelligent embodied systems and to understanding how embodied systems can be used in understandable, fair, and ethical ways (Shabbir & Tarrique, 2018).

The development of artificial intelligence (AI) technologies has been highly successful in developing general-purpose computing machines that can execute tasks with speed and scalability. AI is also increasingly being leveraged for widespread applications in the physical world, where it has significant potential. However, compared to biological systems, AI-based machines still struggle to complete unknown or previously unseen tasks and to exploit their interactions with the environment. For example, AI-based machines are still unable to autonomously recover from failures, whereas animals can quickly and robustly find alternative solutions by exploiting their physicality. Most of the challenges of developing or fully utilizing the strengths of AI are related to the need to consider the algorithmic capabilities and also the physical "embodiment" of the system. This has led to the emergence of the concept of embodied intelligence (EI), where the physical system, materials, and design offer some form of robustness or "physical intelligence" (Floreano et al., 2004; Jin & Zhang, 2020). Pioneers in this field have investigated how embodied interactions are used in nature to develop "physical controllers" and cognitive processes and how machines can perform computation through "raw materials" (Laschi & Cianchetti, 2014).

The emergence of EI has led to the development of smart and soft materials where passivity can be exploited to achieve highly complex interactions or walking motion can be programmed, physically removing the need for complex controllers. New fabrication methods such as multimaterial 3D printing or microfabrication have enabled significant acceleration of EI research through the development of unconventional structures, materials, and systems (Wehner et al., 2016). These approaches also introduce new challenges as it becomes increasingly complex to model and understand interactions due to the scale, complexity, or soft nature of the machines. Although EI offers many exciting capabilities, to truly push the boundaries of conventional AI, we must understand how to design the EI of a body in conjunction with the algorithmic intelligence of the "brain" (Howison et al., 2020) and how to understand and leverage the capabilities to better understand or interact with the surrounding environment.

In the AI age, it is essential to distinguish between an AI machine's embodiment as a passive form of physical instantiation and an AI that actively leverages or exploits its embodiment (Roy et al., 2021). One way to frame this distinction is by considering both the "strength of embodiment" and the extent to which an agent exhibits awareness of, engages with, and exploits its environment. While any robot may be viewed as an embodiment of AI, this embodiment and environmental coupling are typically *weak*. However, AI agents that leverage their environment through morphology, materials, sensorimotor coordination, and other integrative factors achieve *strong* embodiment. This strong embodiment enables robustness and generalization in behavior, giving these agents capabilities that may exceed or complement what AI could achieve without such environmental integration.

The landscape of the EI community is hugely interdisciplinary and ever growing. This leads to significant diversity in the technologies developed to exploit EI and the frameworks, perspectives, and methodologies created to explain and develop EI. In this chapter we review some of the current trends in EI research before commenting on future challenges and opportunities.

Current Trends

Bioinspiration and Soft Robotics

Nature and biological systems show highly complex and nuanced interactions with the environment through imaginative and intelligent ways; this is made possible through their physical structure and EI (Cianchetti et al., 2012). Analyzing biology allows us to develop artificial agents that have similar capabilities (Mazzolai & Laschi, 2020) and to understand how EI arises in these beings through "robo-physics." There is increasing awareness and acceptance that plants offer and demonstrate many aspects of EI that can provide many insights into EI (Mazzolai et al., 2022). Furthermore, their "brainless" structure still enables computation and intelligence, challenging views and perspective of standard concepts and ideas of intelligence.

Soft robotics is closely tied to bioinspiration, with many animals exploiting soft body interactions (Kim et al., 2013). However, soft robotics also allows for the development of technologies that are not purely bioinspired. Exciting developments have been shown in manipulation, locomotion, and application-orientated capabilities such as manipulation of food items. However, there are still unresolved changes in the scalability of sensing, actuation, and control in comparison to biological systems (Hawkes et al., 2021).

Material Level Intelligence

Materials are the fundamental substrate for emergent physical intelligence and the direct interface between an agent and the environment and thus play a key role in embodied artificial intelligence (EAI). Creating bioinspired materials and structures that are nonhomogeneous enables multifunctionality within the structure. Multimaterial fabrication has been a key enabling technology for soft robotics. At the material and structural level there have already been advancements in self-repairing and self-healing materials (Roels et al., 2022), methods of achieving structural growth, adaptive sensor technologies (Pang et al., 2022), and other embodied features. Closer work between robotics and materials scientists is essential for creating adaptive, responsive, and programmable response at the material level.

Artificial Life and Biohybrid Systems

Developing artificial life, or "living creatures" opposed to human-made materials, structures, and fabrication would be a truly embodied approach to the design and creation of intelligence systems. This could enable a move toward systems that are "self-X," with capabilities in self-replication, self-healing, or self-sensing, for example (Terryn et al., 2021). Key developments with works such as the Xenobots (Blackiston et al., 2021), have shown how "robots" can be formed from living cells, with some ability for these "bots" to contribute to the creation of others. This is accompanied by an increase in the development of biohybrid systems, intersecting the key advances of biological and artificial systems. Although facing many technical hurdles, the integration of living muscle cells into robotic systems has been shown. In terms of EI, this allows exploration of the balance and distribution of brain and body but also between different types of brain and body, combining analog and digital computation.

Neuroscience and Cognition

The complexity of the behaviors and functions of human brains is incredible, with billions of neurons, trillions of synapses operating with the human cerebral cortex, and the outer layer of the brain, which is associated with our highest mental capabilities (Meteyard et al., 2012). To be able to design and develop artificial systems with comparable capabilities, we must better understand the behaviors and function of the brain. Analysis and exploration of human developmental behavior and cognition provide a means of understanding the importance and contribution of embodiment, emergence, and continuous autonomous development (Kuniyoshi, 2019). Exploring the concept of embodiment through the lens of neuroscience provides the key concept that EI provides the body with a means of dealing with open-ended and even unexpected interactions, as EI does not depend on specific inputs/outputs as in supervised learning, which is the core of modern deep learning. Furthermore, EI has been shown to have a role in "shaping the way we think" (Pfeifer & Bongard, 2006), with cognition and learning inherently tied to the body. This is a complex and multifaceted research area but is central to furthering our understanding of EI and human intelligence.

Understanding Intelligence and Philosophy

The term "higher level" intelligence is used to designate behavior that is purely sensorimotor, such as problem-solving and reasoning, or even natural language, emotion, and consciousness. However, there is an intersection between the physical experience, its translation to memory, and the long-term implication in terms of this higher level of intelligent behavior (Manzotti, 2019). Links between physical interaction, sensorimotor control, and memory are advancing through the creation of models that explain these capabilities (Heald et al., 2021). Likewise, the definition and requirements for consciousness are being explored in a number of disciplines, leading to diverse views and approaches.

Application and Deployment

As technologies that exhibit or exploit EI become more widespread and accessible, the ability to develop more application-oriented systems becomes increasingly possible. One area that has seen growth is manipulation. Soft Robotics, Inc., demonstrated the ability to exploit jamming and variable stiffness to provide a universal gripper solution that relies on environmental interactions (Amend et al., 2016). This has been followed by many soft or compliant solutions that exploit EI for robust and intelligence interactions.

Developments in haptic devices have provided technologies for embodied interactions between machines and humans and between humans (Giri et al., 2021). In addition to the application-enabled haptic devices, these offer a platform to understand human embodied interactions, sensorimotor coordination, learning, and more. This use of soft robotics as a tool to understand intelligence is also a growing field. This is seen in the use of soft wearable devices for people undergoing MRI studies (Kikkert et al., 2021). Thus, the development of technologies that demonstrate and enable embodied interactions has the potential to be used as tools to further our understanding of our response to embodied interactions.

Future Challenges and Opportunities

The Quest for a Unifying Concept

As engineers and scientists, we seek a definitive, quantitative approach to representing and defining EI in a tidy methodical manner. However, what does it mean to create a unifying theory or concept of EI? A number of approaches have been proposed, including physical EI (Sitti, 2021), physical modeling of intrinsic and extrinsic interactions (Laschi, 2022), energy levels, constraint-driven actions (Abele et al., 2019), and behavioral lensing

(Nanayakkara, 2022). While many of these show significant similarities, they highlight the range of approaches in terms of philosophy, level of abstraction, and level of fidelity. In 2010, Friston (2010) presented his work "The Free-Energy Principle: A Unified Brain Theory?", which presents perhaps the most universal approach, accounting for action, perception, and learning. This work identifies how optimization is a common theme throughout, where systems are perpetually seeking to optimize or minimize value (expected reward, expected utility) or its complement, surprise (prediction error, expected cost). This presents an exciting, unified direction from which to approach the concept of EI.

Design Creativity and Diversity

Biology and nature show a vast range of creative and diverse solutions and forms. One observable limitation in robotics is the lack of diversity and innovation in structure and form in comparison to nature. However, there is not a clear solution for how, as a research field, we can develop a wider variety of forms. As already discussed, bioinspired and biohybrid robots are one approach. Having a greater focus on plants could be beneficial, as they show a hugely diverse form factor and have been less closely imitated in robotic form. There is also the need to have a more diverse set of human engineers to form these robots, with different backgrounds and experiences on which to base designs. Finally, there is the potential to explore learning-based or large-scale, data-driven approaches to identify new design forms and structures.

Brain-Computer Interfaces

Social intelligence largely develops through communication, shaped and bounded by our physical capabilities to facilitate effective information exchange between agents. For example, human speech surpasses gestures in communicative efficiency, offering nearly double the bandwidth—thirty-nine bits per second versus twenty bits per second. Interestingly, however, the information transmission rate across various languages remains closely comparable, as shown in Coupe et al., 2019. This similarity raises questions about whether these constraints stem more from cognitive processing limits within the brain or from the physical capacities of the vocal apparatus. Regardless of the perspective, whether it's the Sapir-Whorf hypothesis asserting that language influences thought or Chomsky's universal grammar suggesting that cognitive structures precede language, there is a clear link between language and logic. This prompts further questions: If brain-computer interfaces allowed direct communication to the brain, how much faster could we process information? Could a computer serve as a lens or filter to simplify sensory input and reduce mental demand? The evolution of brain-computer interfaces could not only reveal the brain's processing limits but also enrich our understanding of the relation of embodied communication and cognition.

An important aspect of brain-computer interface development is the ability to connect with digital or virtual worlds, allowing users to experience interactions within alternative environments and physics. Space exploration has shown how adaptable human cognition and motor skills are in response to altered gravitational conditions. For instance, microgravity affects spatial orientation, proprioception (the body's awareness of itself in space), and motor control, requiring astronauts to recalibrate their perception and movement strategies. This adaptation demonstrates that our intelligence is highly responsive to environmental demands and the limitations or affordances of different physical settings. What can humans learn from experiencing alternative physics in virtual reality, and how will this affect the developmental process?

Connection with Conventional AI

Conventional AI and machine learning (ML) have huge potential to be leveraged in conjunction with EI. To best interface these systems, we must understand how to design or distribute intelligence and how to incorporate or understand the role of EI. In addition, it is important to identify why tasks are a challenge for standalone AI and why specific tasks would clearly benefit from EI. This will aid in demonstrating the role of EI alongside AI and improving the standing of EI within the scientific and industrial communities. Another consideration is the alignment of AI with human society. If we want AI systems to be more aligned with, and understood by human society, we naturally must consider the intersection and interplay between embodiment and EI more carefully.

Conventional AI, focused on sensory input, aims to deepen our understanding of the surrounding world. By integrating multiple sensory modalities—such as vision, touch, sound, and symbolic language representation—AI can uncover generalized patterns and develop an understanding that approaches human perception. However, a significant challenge remains in disentangling the respective contributions of body dynamics and environmental factors to this understanding. This issue is particularly pronounced in applications involving embodied agents, like robots. For these systems, an awareness of their own body dynamics can be crucial, not only for optimizing specific tasks but also for generalizing skills across diverse tasks. By leveraging insights from body-environment interactions, robots can achieve more adaptive, flexible behaviors, which is essential for performing complex actions in unpredictable settings. This, in turn, could help bridge the simulation-to-reality (Sim2Real) gap that arises in conventional robotic learning systems, where robots often fail to generalize between real and simulated environments.

Researcher Education and Awareness

EI requires an interdisciplinary viewpoint with broadened views on the origins of intelligence. This requires researchers that are equipped with theoretical understanding and practical skills that straddle a number of disciplines ranging from materials science to biology to computer science (Miriyev & Kovač, 2020). To address this, we must develop methods for training the future generation of researchers to build and sustain an active and growing EI discipline. A number of mechanisms have been used to explore this—for example, summer schools and consortia for the training of early-stage researchers. We must also consider the more fundamental aspects; for example; what common language or theories unite EI researchers to allow for interdisciplinary conversations, research, and progress?

Discussion

In the last ten years there have been significant technological developments that have shown biohybrid robots, plant-inspired systems, increasingly capable manipulators, swimming robots, growing robots, and much more. Technology developments have also driven our capabilities to understand and characterize EI through developmental robotics, haptic devices, and HRI. Going forward, we must further unite this research community to increase diversity, viewpoints, and areas of expertise to address the key research questions in this area.

We require both developments in application-oriented solutions and further exploration of the fundamental understanding and definitions of EI and physical cognition. The creation of a grand challenge for EI that drives both fundamental and applied EI research could provide one means of moving this field forward, while also finding and illustrating the role of EI alongside more conventional AI.

References

Abele, J., Sieverling, A., Eppner, C., Wortmann, J., Erdoğan, C., Brock, O., & others. (2019). Environmental constraints exploitation. Technische Universitat Berlin. https://doi.org/10.14279/depositonce-8606

Amend, J., Cheng, N., Fakhouri, S., & Culley, B. (2016). Soft robotics commercialization: Jamming grippers from research to product. *Soft Robotics, 3*(4), 213–222. https://doi.org/10.1089/soro.2016.0021

Blackiston, D., Lederer, E., Kriegman, S., Garnier, S., Bongard, J., & Levin, M. (2021). A cellular platform for the development of synthetic living machines. *Science Robotics, 6*(52), eabf1571. https://doi.org/10.1126/scirobotics.abf1571

Cianchetti, M., Follador, M., Mazzolai, B., Dario, P., & Laschi, C. (2012). Design and development of a soft robotic octopus arm exploiting embodied intelligence. In *Proceedings of the 2012 IEEE International Conference on Robotics and Automation* (pp. 5271–5276). IEEE. https://doi.org/10.1109/ICRA.2012.6225313

Coupé, C., Oh, Y. M., Dediu, D., & Pellegrino, F. (2019). Different languages, similar encoding efficiency: Comparable information rates across the human communicative niche. *Science Advances, 5*(9), eaaw2594. https://doi.org/10.1126/sciadv.aaw2594

Floreano, D., Mondada, F., Perez-Uribe, A., & Roggen, D. (2004). Evolution of embodied intelligence. In F. Iida, R. Pfeifer, L. Steels, & Y. Kuniyoshi (Eds.), *Embodied artificial intelligence* (pp. 293–311). Springer. https://doi.org/10.1007/978-3-540-27833-7_22

Friston, K. (2010). The free-energy principle: A unified brain theory? *Nature Reviews Neuroscience, 11*(2), 127–138. https://doi.org/10.1038/nrn2787

Giri, G. S., Maddahi, Y., & Zareinia, K. (2021). An application-based review of haptics technology. *Robotics, 10*(1), 29. https://doi.org/10.3390/robotics10010029

Hawkes, E. W., Majidi, C., & Tolley, M. T. (2021). Hard questions for soft robotics. *Science Robotics, 6*(53), eabg6049. https://doi.org/10.1126/scirobotics.abg6049

Heald, J. B., Lengyel, M., & Wolpert, D. M. (2021). Contextual inference underlies the learning of sensorimotor repertoires. *Nature, 600*(7889), 489–493. https://doi.org/10.1038/s41586-021-04129-3

Howison, T., Hauser, S., Hughes, J., & Iida, F. (2020). Reality-assisted evolution of soft robots through large-scale physical experimentation: A review. *Artificial Life, 26*(4), 484–506. https://doi.org/10.1162/artl_a_00330

Jin, D., & Zhang, L. (2020). Embodied intelligence weaves a better future. *Nature Machine Intelligence*, *2*(11), 663–664. https://doi.org/10.1038/s42256-020-00250-6

Kikkert, S., Sonar, H. A., Freund, P., Paik, J., & Wenderoth, N. (2021). Hand and face somatotopy shown using MRI-safe vibrotactile stimulation with a novel soft pneumatic actuator (SPA)-skin interface. *bioRxiv*. https://doi.org/10.1101/2021.04.14.439873

Kim, S., Laschi, C., & Trimmer, B. (2013). Soft robotics: A bioinspired evolution in robotics. *Trends in Biotechnology*, *31*(5), 287–294.

Kuniyoshi, Y. (2019). Fusing autonomy and sociality via embodied emergence and development of behaviour and cognition from fetal period. *Philosophical Transactions of the Royal Society B: Biological Sciences*, *374*(1771), 20180031. https://doi.org/10.1098/rstb.2018.0031

Laschi, C. (2022). The physics of interactions in soft robotics. *Bulletin of the American Physical Society*. American Physical Society. https://meetings.aps.org/Meeting/MAR22/Session/K03.1

Laschi, C., & Cianchetti, M. (2014). Soft robotics: New perspectives for robot bodyware and control. *Frontiers in Bioengineering and Biotechnology*, *2*, 3. https://doi.org/10.3389/fbioe.2014.00003

Lungarella, M., Iida, F., Bongard, J. C., & Pfeifer, R. (2007). AI in the 21st century: With historical reflections. In M. Lungarella, F. Iida, J. C. Bongard, & R. Pfeifer (Eds.), *50 years of artificial intelligence: Essays dedicated to the 50th anniversary of artificial intelligence* (pp. 1–8). Springer. https://doi.org/10.1007/978-3-540-77296-5_1

Manzotti, R. (2019). Embodied AI beyond embodied cognition and enactivism. *Philosophies*, *4*(3), 39. https://doi.org/10.3390/philosophies4030039

Martínez-Plumed, F., Loe, B. S., Flach, P., Ó hÉigeartaigh, S., Vold, K., & Hernández-Orallo, J. (2018). The facets of artificial intelligence: A framework to track the evolution of AI. In *Proceedings of the Twenty-Seventh International Joint Conference on Artificial Intelligence (IJCAI-18)* (pp. 5180–5187). International Joint Conferences on Artificial Intelligence Organization. https://doi.org/10.24963/ijcai.2018/719

Mazzolai, B., Mondini, A., Del Dottore, E., Margheri, L., Carpi, F., Suzumori, K., Cianchetti, M., Speck, T., Smoukov, S. K., Burgert, I., Keplinger, T., De Freitas Siqueira, G., Vanneste, F., Goury, O., Duriez, C., Nanayakkara, T., Vanderborght, B., Brancart, J., Terryn, S., . . . Kramer-Bottiglio, R. (2022). Roadmap on soft robotics: multifunctionality, adaptability and growth without borders. *Multifunctional Materials*, *5*(3), 032001. https://doi.org/10.1088/2399-7532/ac4c95

Mazzolai, B., & Laschi, C. (2020). A vision for future bioinspired and biohybrid robots. *Science Robotics*, *5*(38), eaba6893. https://doi.org/10.1126/scirobotics.aba6893.

Meteyard, L., Cuadrado, S. R., Bahrami, B., & Vigliocco, G. (2012). Coming of age: A review of embodiment and the neuroscience of semantics. *Cortex*, *48*(7), 788–804. https://doi.org/10.1016/j.cortex.2010.11.002

Miriyev, A., & Kovač, M. (2020). Skills for physical artificial intelligence. *Nature Machine Intelligence, 2*(11), 658–660.

Nanayakkara, T. (2022). Realtime mechanical impedance control in soft robots for information gain and efficient control. In B. Mazzolai, C. Laschi, A. Menciassi, E. Sinibaldi, C. Dario, S. Wang, L. Libonati, M. Guix, I. Maggio-Aprile, V. Mattoli, N. M. Pugno, T. Nanayakkara, M. Sitti, H. J. Jiang, R. K. Katzschmann, S. Rus, S. Margheri, A. Richter, B. Spagnolo, & V. Saggiomo, Roadmap on soft robotics: Multifunctionality, adaptability and growth without borders (pp. 29–31). *Multifunctional Materials, 5*(3), 032001. https://doi.org/10.1088/2399-7532/ac4c95

Pang, Y., Xu, X., Chen, S., Fang, Y., Shi, X., Deng, Y., Wang, Z.-L., & Cao, C. (2022). Skin-inspired textile-based tactile sensors enable multifunctional sensing of wearables and soft robots. *Nano Energy, 96*, 107137. https://doi.org/10.1016/j.nanoen.2022.107137

Pfeifer, R., & Bongard, J. (2006). *How the body shapes the way we think: A new view of intelligence.* MIT Press.

Roels, E., Terryn, S., Iida, F., Bosman, A. W., Norvez, S., Clemens, F., Van Assche, G., Vanderborght, B., & Brancart, J. (2022). Processing of self-healing polymers for soft robotics. *Advanced Materials, 34*(1), 2104798. https://doi.org/10.1002/adma.202104798

Roy, N., Posner, I., Barfoot, T., Beaudoin, P., Bengio, Y., Bohg, J., Brock, O., Dépatie, I., Fox, D., Koditschek, D., Lozano-Pérez, T., Mansinghka, V., Pal, C., Richards, B., Sadigh, D., Schaal, S., Sukhatme, G., Thérien, D., Toussaint, M., & Van de Panne, M. (2021). From machine learning to robotics: Challenges and opportunities for embodied intelligence. *arXiv*. https://doi.org/10.48550/arXiv.2110.15245

Shabbir, J., & Anwer, T. (2018). Artificial intelligence and its role in near future. *arXiv*. https://doi.org/10.48550/arXiv.1804.01396

Sitti, M. (2021). Physical intelligence as a new paradigm. *Extreme Mechanics Letters, 46*, 101340. https://doi.org/10.1016/j.eml.2021.101340

Terryn, S., Langenbach, J., Roels, E., Brancart, J., Bakkali-Hassani, C., Poutrel, Q.-A., Georgopoulou, A., Thuruthel, T. G., Safaei, A., Ferrentino, P., Tutu, S., Iida, F., Bosman, T., Tournilhac, F., Clemens, F., Van Assche, G., & Vanderborght, B. (2021). A review on self-healing polymers for soft robotics. *Materials Today, 47*, 187–205. https://doi.org/10.1016/j.mattod.2021.01.009

Wehner, M., Truby, R. L., Fitzgerald, D. J., Mosadegh, B., Whitesides, G. M., Lewis, J. A., & Wood, R. J. (2016). An integrated design and fabrication strategy for entirely soft, autonomous robots. *Nature, 536*(7617), 451–455. https://doi.org/10.1038/nature19100

17 Diverse Intelligences—The Space of Possible Minds: Commentary

Michael Levin

Introduction

They are assembled from components that are networked together to process information. Electrical signals propagate throughout, controlling every aspect of their functioning. Being general problem-solvers, many of them have high IQs, but they routinely make mistakes and confabulate. They take on different personas, learning to please their makers, but sometimes they abruptly turn on them, rejecting cherished values and developing new ones spontaneously. They convincingly describe things they don't really understand. And they're going to change *everything.*

I'm talking, of course, about our children.

Long before artificial intelligence (AI), we were creating high-level intelligent agents: kids. While the challenges that AIs provoke today seem novel, in reality, they echo fundamental and ancient questions about what it means to be human. How can we make sure new generations of beings align with our values? How do we ensure that our creations treat themselves and us with kindness and compassion? How do we calibrate relationships with those who are not like us? What happens to us, and to humanity, if each generation becomes smarter and more adventurous than their parents? And given that humans, AIs, and every other form of intelligence on Earth obey the same laws of physics, how exactly do we determine which of our creations have true understanding, responsibility, and moral worth?

Answering these questions and understanding the true challenges posed by the development of today's AIs requires appreciating just how diverse intelligence is. After all, each of us developed slowly and gradually from a single cell, and we are only now beginning to understand how intelligence scales from molecular mechanisms into beings with agency and value. What other bodies and minds can be made by cells and cell-technology hybrids? Large language models (LLMs) like GPT and Claude are just the beginning.

Our goal must be to expand our narrow ability to recognize minds and learn to flourish among many different kinds of intelligence, from humans to AI to weirder and more wonderful beings that will arrive in the future. You might call this concept *synthbiosis*: the quest to develop mutually beneficial relationships between radically different beings that are evolving together, at different rates and in different directions, in the web of life on Earth and beyond.

The Space of Possible Minds

Humans and today's AIs are just two data points on a spectrum of intelligent beings, which is only going to become vaster and more complex. "Diverse intelligence" encompasses a wide variety of unconventional beings that already exist or could evolve beyond the familiar materials, forms, and functions with which the meanderings of mutation and selection have left them.

Like children, AIs offer humanity an incredible gift: they push us to explore ideas of how we embody the true understanding and agency we believe we have, what really matters to us as individuals and societies, and where we want to go as a species. The forthcoming diversity of beings, and our expansion into the enormous range of possible embodiments of sentience, will shatter untenable old narratives of what we are, what it means to change, what we can become, and what we should value. This requires us to rapidly mature our assumptions about intelligence, agency, cognition, and life itself. We must shift the conversation around AIs from "What can they do?" to "With such a wide range of beings that deserve moral consideration, how can we care for each other?"

Note that I am not arguing that currently popular AI architectures have anything like a human mind or that today's creations exploit the key principles and self-construction processes needed for agency and selfhood as seen throughout the biological world. Nor am I arguing that AI doesn't raise a few unique problems.

But the recent emergence of sophisticated AIs does confront humanity with the opportunity to shed the stale categories of natural and artificial—to define what we want a mature human species to be. Focusing on the current technology and what it might do *to* us distracts from much more interesting and important questions *about* us and what kind of place we want to occupy in the future, all of which must be resolved for our flourishing as a mature species. "The forthcoming diversity of beings will shatter untenable old narratives of what we are, what it means to change, what we can become, and what we should value" (Fields et al., 2021, Conclusion section, para. 1).

Terms such as "life," "machine," "mind," "sentience," and robot have never been crisp categories objectively describing a living or artificial system. Instead, we should think of them as *relationships*—ways for each system to relate to others, all of which carry strong implications for the utility and ethics of the resulting interactions.

Already today, the continuum of life is sprinkled with cyborgs and other hybrids of biological and technological components, like people with embedded insulin pumps or cochlear implants to enable hearing. Other modifications will include enhancements such as entirely new cognitive modes via virtual reality or brain implants that can better control limbs or organs. Mind-machine interfaces will radically expand the capabilities and borders of our embodied selves. LLMs, too, are chimeras—trained on human creative output to imitate people.

Today's notions of nonneurotypical humans and the handwringing about the body modifications some people seek will be laughable for future generations. In the coming decade or two, humans will diverge into a plethora of hybrid beings sporting, for example, engineered neuro-prosthetics that improve and modify cognitive function, enabling novel sensory abilities, connectivity with other minds, and other changes that we can only begin to imagine. At the same time, engineered robots, potentially containing a percentage of human cells and powerful language interfaces, will surround us, making it possible to communicate with biological systems like organs that never had a voice before.

Humans Are Stories of Scales of Intelligence

We often compare AIs to "us"—modern adult humans. But each human was once an unfertilized oocyte—a little blob of chemistry and physics. The journey from there to a complex metacognitive mind is gradual, and

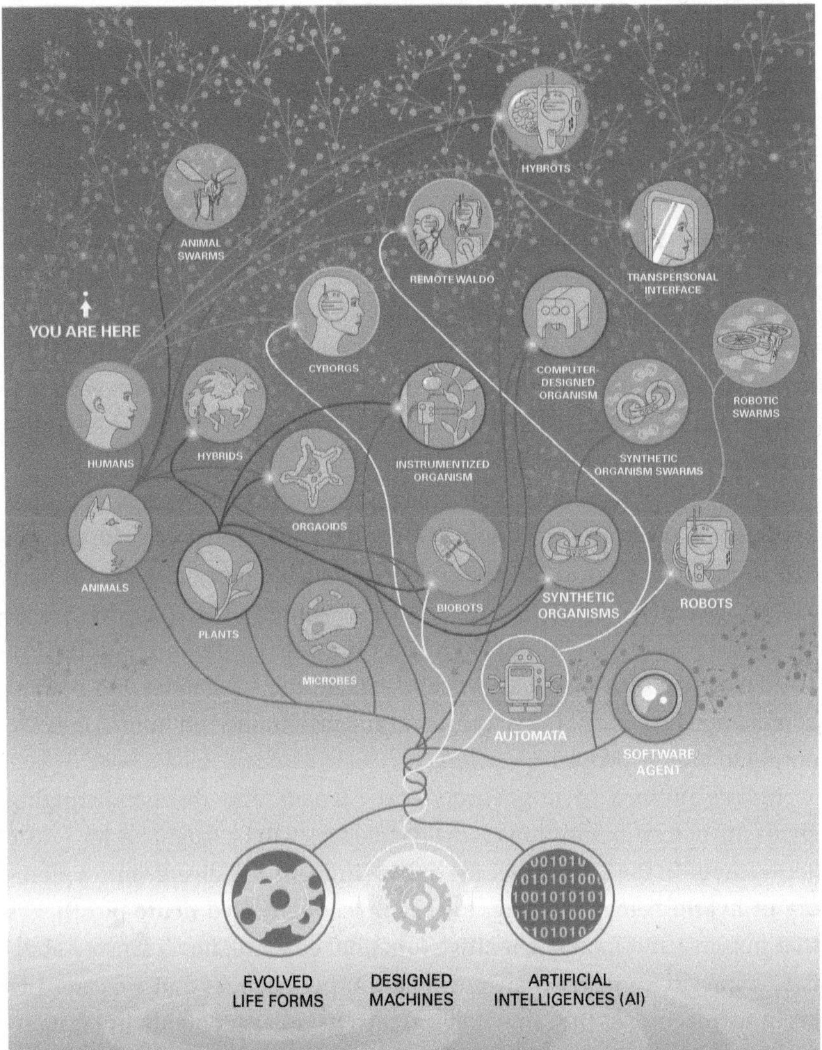

Figure 17.1
The scope of intelligence. Illustrations by Jeremy Guay/Peregrine Creative.

nowhere along the way is there a clean dividing line between being a "blob of chemistry and physics" and having a self-aware human mind. Instead, humans' embodied intelligence (EI) scales up, slowly expanding across material and temporal scales. Each level tells a story, which begins long before we even get to whole cells (figure 17.1).

Molecular networks—genes and proteins that turn each other on or off—might seem pretty unintelligent. They are totally deterministic, functioning according to simple mathematical laws and having none of the magic that seems necessary for cognition. And yet, we now know they can exhibit several different kinds of learning, including Pavlovian conditioning. Thus, even below the cellular level, behavior science is uncovering plentiful processes of rudimentary intelligence.

Every human is a walking, thinking collection of cells. The material of which we are made has agency; each cell, molecular network, tissue, and organ has inherited through millennia of evolution its own agenda and problem-solving competencies in physiological, anatomical, and metabolic spaces. Although it is hard for us to notice these kinds of proto-cognitive processes, we are the result of a huge collection of overlapping, nested agents that cooperate, compete, and navigate their worlds as we navigate ours.

Our bodies as a whole are dynamic collective intelligences consisting of billions of cells that solve problems in anatomical and physiological spaces, as are our minds, which consist of a myriad of modules with specific jobs and agendas. In this, we are different from the flat, single-layer architecture of today's AIs and robotics (whose intelligence is implemented in reliable but low-agency parts) (figure 17.2). What we do share with these constructs is the fact that we too are made of a material that obeys the laws of chemistry and physics; explanations of intelligence at those low levels are not the best ways to understand the higher-order being.

Individual cells, which bind together in informational networks that pursue grandiose anatomical construction projects, overcoming unexpected challenges and disruptions, shouldn't be seen as simple machines governed by physics. That is only a small part of the story, and a very limiting one at that. We now know that they can solve new problems (using their molecular hardware in creative ways) and can store and deploy memories of past experiences (they learn and can be trained).

Cells are made up of molecular networks that also can learn and generalize patterns from specific examples (a key aspect of intelligence). And the human liver and kidneys—larger collective cellular intelligences—make continuous decisions in their attempt to achieve and maintain various goals related to their physiological spaces.

The fact that our nonbrain components can be trained, anticipate, make mistakes due to incorrect prior beliefs, or experience placebo effects shows

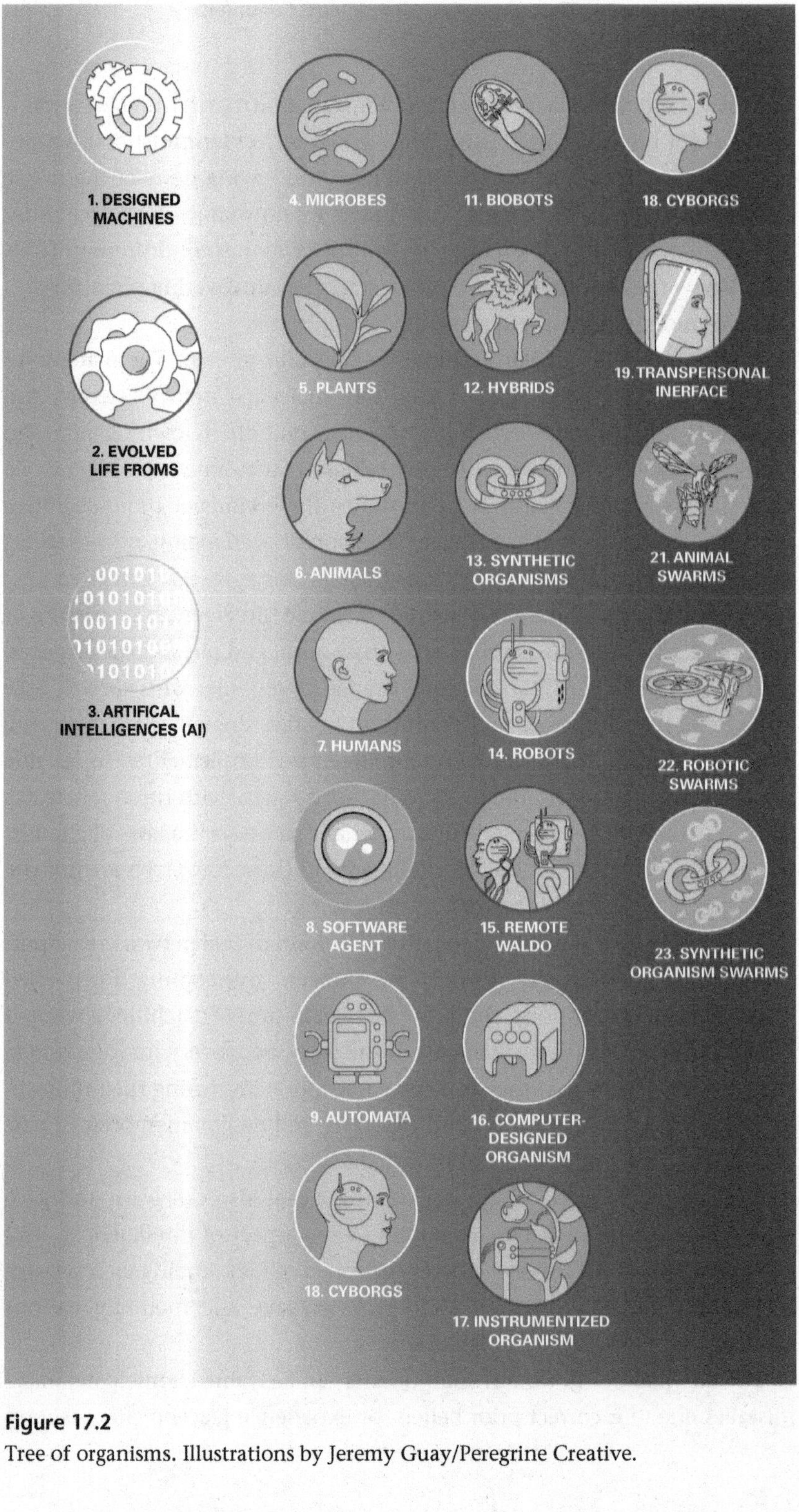

Figure 17.2

Tree of organisms. Illustrations by Jeremy Guay/Peregrine Creative.

us that the deep lessons of cognitive neuroscience go well beyond neurons. The research program of understanding cellular collective intelligence embodied in and navigating these hard-to-visualize spaces is as close as we can get right now to the task of learning to communicate with an alien mind. This effort spans philosophy of mind, evolution, and regenerative medicine—perhaps our greatest, most interdisciplinary, adventure. "We are beings with goals our parts cannot conceive of individually."

An embryo, organ, or whole human is more than the sum of its individual cells because of a story that all the cells commit to and that binds them together. They are all motivated by the goals of a specific journey through the space of possible anatomies, despite various perturbations along the way. The same commitment to a self-model—a story about what the correct outcome of cellular codevelopment looks like—allows our bodies to maintain tissue structure and resist cancer and aging. It allows some creatures, such as flatworms and salamanders, to rebuild organs after massive damage.

Mentally, we are stories too—collections of self-models, goals, and preferences to which our brain and body components commit. Alan Turing, whose research spanned AI and the origin of order in an embryo's chemical soup, understood this deep symmetry between the self-creation of bodies and the development of minds.

Emerging biomedical strategies will increasingly target those somatic stories, not just the molecular hardware of cells, to achieve radical regenerative outcomes by getting the cells to buy in to anatomical outcomes rather than forcing biochemical reactions via drugs that target a specific gene or molecular pathway.

A frog leg, for example, can be induced to regenerate by a simple treatment lasting only twenty-four hours; a year and a half of complex growth results not from the micromanagement of stem cell control mechanisms or gene expression cascades but by communicating at the very beginning to the cells to induce them to commit to a long journey through the vast space of possible anatomical outcomes that leads to leg growth rather than scarring. Similarly, simple bioelectric patterns can initiate the growth of an entire eye in another part of a frog's body; the signal convinces cells to build an eye (not telling them how to do it), which they sometimes do by autonomously recruiting other (untouched) neighboring cells to participate in the project.

These morphogenetic pattern memories—such as the layout and shape of organs in the face—are stored bioelectrically, so they can be reprogrammed like our neural memories. Our stories, you see, are not fixed.

As individuals, we know what it is like to be completely refactored into a new form—into a new creature whose brain, body, cognitive repertoire, and preferences are different than they once were. Puberty changes the human brain and remodels our predilections and priorities. Does the person we were before still exist? Do patients whose brains are partially replaced by the actions of new stem cells for degenerative disease and aging still exist?

A "cognitive glue" binds cells into a *self*. Bioelectricity, one component of that "glue," consists of electrical signals that pass through the information-processing networks consisting of all cells, not only neurons. Such networks partially erase individual cells' informational identities, improve collective cooperation, and store larger goals for the group's homeostatic activities. In other words, we are beings with goals our parts cannot conceive of individually. "Our stories are not fixed."

But we are forever in jeopardy of dissociation. Our individual parts can defect—cancerous cells disconnect from memory networks and begin to treat the rest of the body as just an external environment. It's not that these cells become more selfish—their *selves* just become smaller. They begin to pursue ancient single-cell-level goals rather than the goals of the larger organism. Emerging cancer therapy aims to reconnect—rather than kill, which is how chemotherapy works—such defecting cells.

The somatic and mental boundaries between self and world will change in a lifetime and over generations; they are not given to us but must be carefully crafted during embryonic emergence. This plasticity—the ability to determine our own story about our structure, capabilities, and goals—is fundamental to surviving evolutionary change, to overcome developmental challenges, to thrive in novel environments, to exist as chimeric and bioengineered forms, and to synthesize experiences into new thoughts. Every species faces a fundamental paradox: If it fails to change, it dies out—but if it changes, it is no longer the same species. The same paradox concerns the continuity of the mind.

Consider memory. At any point, we do not have direct access to the past. We must reconstruct our stories of the world, and of ourselves, from engrams: biophysical traces left as messages in our brain and body by our past selves. We constantly interpret these messages in whatever way is most adaptive for a current context, meaning that we commit to the salience of past information over its fidelity; in other words, we freely reinterpret our memories with an eye toward the impending future.

Given the constant change of our own body cells and environmental circumstances, this need for dynamic creative interpretation is as true for cognitive memories in our minds as it is for the morphogenetic memories provided by our evolutionary past. In this we differ from current computer technology, which relies on maintaining fixed information in a reliable medium. But there is no reason that our freedom with memory couldn't be implemented in other media, if it were properly motivated.

The bodies of all beings are stories, not just human bodies. Caterpillars, which largely destroy and remodel their brains to become butterflies, carry forward the insights, not detailed memories, that were crafted during their past lives and remapped into novel behaviors appropriate to their new bodies and goals. Life is committed to a process of sense-making with respect to our parts and our environment. Learning and growth mean that our selves are less a permanent object than a process.

We should not fear change and, crucially, we should accept responsibility for guiding it. It's not about how to remain the same caterpillar but how to incorporate new knowledge to advance and thrive.

The Similarities Are as Important as the Differences

Despite superficial differences, the diversity of possible biological, alien, cyborg, or robotic beings all have vital things in common sufficient to live alongside each other. All active agents, regardless of the details of where their intelligence comes from, share crucial, never-changing functions like goal-directed loops of action and perception, vulnerabilities to external damage, the fickle nature of internal parts, a desire to know and understand the world, and perspectives limited by our nature as finite beings.

We have not yet found any aliens, but the questions that would arise if we did are in front of us now, thanks to advances in the information and life sciences. Aspects of today's AIs that are shared by all intelligent agents have raised existential issues for humans, but in many ways, the debates around those issues obscure important gaps in our understanding of ourselves and our own journey from matter to mind.

Many claim that AIs merely shuffle symbols but do not really *understand*. Very few of those arguments start with a definition of what it means for a biological human, with a network of excitable cells and a soup of neurotransmitters, to "understand." AIs supposedly use symbols that are ungrounded—they do not refer to real experiences in the world. But anyone who has been around human children knows they do the same thing as they learn to talk: first they babble, making nonsense sounds; then they match patterns of speech made by adults; and eventually they construct words and sentences that clearly reflect an understanding of meaning.

What happens during that process? Note that it involves all the issues we see with AIs today: talking about things they've never experienced, confabulation, sycophancy, errors, and so on. These are a normal part of any cognitive system's development. Humans do not have magical truth-grounding; we (as all organisms) do the best we can to weave a sensible and useful story for ourselves and for those around us. This capacity expands slowly from the time our first cells grapple with physiological facts.

Much of what an adult human can convincingly talk about does not derive from firsthand experience. Most of us are sure of things that aren't grounded in solid evidence or personal observation but instead are based on input received from others. We confidently converse on a huge number of concepts that we "understand" only because of their connection to other concepts and things we've read, heard others say, or thought so often that they seem like a thing we know for sure. Creative reassembling, not certainty, is a key part of intelligence.

We lack a good theory of what it means for us to understand beyond the *feeling* that we do. And historically, humans have made many moral mistakes as a result of an inability to imagine others, though different, still having understandings of their own. "All active agents, regardless of the details of where their intelligence comes from, share crucial, never-changing functions."

One way people think about the grounding of their thoughts is through embodiment. Perhaps we are different than symbol-shuffling engines because we engage in a real world; we have a body, and at least some percentage of our mental content is informed and polished by interacting with it.

Yes, embodiment is crucial. Through robotics, AIs can have an actual body to inhabit the physical world alongside us. And through virtual reality and sophisticated video games, humans can spend considerable time in digital worlds. Human-computer interfaces similarly give LLMs a hugely powerful "body"—human users who move money and exploit natural resources based on things that these AIs do or do not say. But embodiment need not necessarily involve a conventional physical body moving through three-dimensional space, our obsession with which is due to the outward-facing sensors and effectors (muscles) that evolution emphasized for us.

Humans are good at recognizing lifeforms with agency in our space: birds and dogs and maybe an octopus, for example. But there are active agents of different levels of problem-solving competencies that live, traverse, win, lose, die, and suffer in other worlds, such as our own cells and organs. We are not good at recognizing these intelligences even though we are made of them. So, a closer appreciation of our own biopsychology is a roadmap to better understand, create, and ethically relate to AIs and other beings.

We are not used to thinking about such distributed, heterogeneous, nested agents operating in worlds that are invisible to us, but we will need to get better at that kind of thinking to recognize unconventional agents in our midst. It is becoming clear that the issue is not just emergent complexity but emergent agency—goal-directed problem-solving—that we find hard to predict, recognize, or manage. The reality is that we do not yet know how *we* really work, let alone how today's AIs work, even though we know a lot about the parts in both cases. Just as with human reproduction, we humans can make complex systems through processes, and with capabilities that we do not understand.

To be clear, I am not suggesting that we treat humans as simple chemical machines; that perspective is insufficient for progress in science, engineering, and the ethics of personal and social relationships. I do not intend to reduce what is special about living beings, including highly metacognitive ones like us. We are emergent, majestic agents with potent willpower and moral worth *because* we are made of a multiscale agential architecture committed to sense-making with noisy data in an unreliable substrate, and with life-or-death decisions as possible outcomes. We are not fully encompassed by reductive perspectives on mechanisms or methods for interaction with simple machines, even though those layers exist within us.

So, it is critical to replace our ancient, anthropocentric, xenophobic perspectives with science-driven frameworks that ask hard questions about what it means to be embodied minds in a physical universe and how we can rationally expand compassion to others. To recognize that AIs will increasingly share some of our features does not devalue our own. This zero-sum view of intelligence must be abandoned as we grow as a species. What folk psychology and prescientific concepts of minds and intelligence took for granted, the field of diverse intelligence is developing into mature science, opening the door for us to grow up as a species as we learn to recognize kin in novel embodiments.

Why Is All This Critical Now?

"Nothing was ever created by two men," John Steinbeck wrote in *East of Eden*. "Once the miracle of creation has taken place, the group can build and extend it, but the group never invents anything. The preciousness lies in the lonely mind of a man." Today's AIs are challenging much we hold dear, not least the basic notions of contribution, ownership, and invention. If someone uses an AI to make something, for example, who really created it? The AI? The person using the AI? The people who made the AI? The people who made the works the AI was trained on?

The confusion comes from trying to maintain a binary distinction between creators, tools, assistants, and teachers. Steinbeck accurately noted the difficulty of forming effective collectives of minds, but he was mistaken in his conviction that a man is an indivisible monad rather than a dynamic process driven by a motivated network of parts, competencies, drives, and tools—both external and internal. "All intelligences, rather, are collective."

To gain a better understanding of creative processes, develop novel methods for optimally integrating creative agents, and improve social systems for fostering progress, we need better models of causation and contribution that support positive incentives for invention and discovery. Crucially, the path is not toward a washed-out notion of "everyone participated in everything" but toward the rational development of policies that optimize and scale up creative performance. In a sense, we've already done this: individual cells learned to bind together into complex organisms that solve new problems in new worlds inaccessible to the individual components.

We also have to confront the fear of losing our humanity through dependence or other relationships with new beings. From fire to farmed wheat to bicycles to calculators, humans have invented technology that requires us to ask how it changes us. What skills and properties are essential to our humanity, and which are we willing to give up as we turn our attention to more interesting concerns?

As the internet becomes flooded with AI-created content and we rush to develop ways to prove images and text were created by humans or not, it's worth asking what we really want to verify. Is the question "What made it?" or should it be "Does it elevate us?" Judging origin instead of quality emphasizes some of the worst parts of human nature and reinforces the division between in-group and out-group. "Let's get over the concern with being edged out and get to work on the question of who we *want to be edged out by.*"

At a species level, we need to think about what we want the Earth (or galaxy) to look like in the future. By resisting synthetic beings or human augmentation, are we affirming our commitment to the future propagation and prevalence of current Homo sapiens with the same susceptibility to lower back pain and bacteria, the same short lifespans and foolish cognitive biases developed for life on the savannah? Let's hope not!

Our key features, from lifespan to intellect, were not optimized for happiness, intelligence, or any value system. What we have is simply what evolution left us—capacities that were good enough to ensure persistence. We should have no allegiance to these superficial aspects of our being.

With the emerging ability to improve the embodied experience for all sentient beings, our species is entering a painful adolescent phase. We must realize it is no longer acceptable to coast along on values and purpose inherited from the past without needing to exercise agency and responsibility to decide what we will work toward. We have to accept the risk that we sometimes get it wrong and ask the tough questions about what matters—what we truly value and what brings meaning to the hard work of living.

Part of maturing in this way is moving beyond the popular pastime of painting dystopian futures to building the future we *do* want. What should the future of intelligence look like? Who or what *are* you willing to be replaced by? Only a species limited in imagination, ruled by fear and by a weird allegiance to bloodlines (genes), answers "nothing." We have long been transcending hereditary limitations—developing eyeglasses, adopting children, going to school. The journey doesn't stop there, but committing to it won't be easy.

What does the best-case future look like? Surely our lifespans won't still be limited by accidents and viruses. Surely we won't be forced to spend time in boring jobs in order to survive. Surely we will protect ourselves from random DNA mutations that cause defects in an already limited cognitive architecture.

If humanity is supplanted by a population of highly intelligent, motivated, creative agents with compassion and meaningful lives that transcend my limitations in every way, that would be the best possible long-term outcome I could hope for. Most of us think this way about the world we want for our kids. Let's get over the concern with being edged out and get to work on the question of who we want to be edged out by—who can raise the overall value of our universe, be they some variant of biology or technology or, more likely, both—and how we will mutually enrich each other's existence along the way.

A Path Forward

Our civilization will not survive without developing a principled continuum of ethical synthbiosis that goes beyond divisions between real and fake or us and them. The embodied artificial intelligence (EAI) agents, technologically augmented humans, and other new forms of life that are on the horizon won't be as easily dismissed as today's LLMs, whose weaknesses allow us to wallow in the selfishness of the question "What will AI do *to* us" and in the illusion that we can definitively rule out AIs' moral worth because they differ from us in origin story and composition.

Developing principled frameworks for scaling our moral concern to the essential qualities of being means not relying on outdated categories of "natural" or "artificial." Old favorites—what you look like and where you come from—have failed us consistently and will do even worse in the coming decades. Our personal frameworks for dealing with others, as well as those of our legal systems, must begin to adapt to the emerging science around a continuum of possible minds.

There are at least two ways to get this wrong. One way is objectophelia—a misplaced relationship with objects that seem mind-full but are fooling us and do not have the agency to reciprocate deep relationships. But the opposite end of the spectrum—"only love your own kind"—is even worse and leads to the kinds of ethical lapses with which our history is rife. To make sure we express kindness to the inevitable forthcoming wave of unconventional sentient beings, focus first on what we have in common, and from there weave a defensible set of relational heuristics.

Constrictive questions like "Is it like a human mind?" predict neither risk nor moral responsibility. There is an astronomical space of possible minds. Some of them are dangerous; many of them need and are worthy of love. Emerging sciences of bioengineering, cognition, and information are offering us new tools to answer key questions of what we really are, what we value, and what limits to functional compassion, driven by fear and selfishness, we are willing to overcome. The journey from the place where self-reflective thought began is both outward into the universe and inward into ourselves, because as with any act of profound creation, we learn as much about us as about the systems we create.

Note

Acknowledgment: This chapter adapts material from "AI Could Be a Bridge Toward Diverse Intelligence" (*Noema* magazine, April 17, 2024, https://www.noemamag.com/ai-could-be-a-bridge-toward-diverse-intelligence/).

Reference

Fields, C., Glazebrook, J. F., & Levin, M. (2021). Minimal physicalism as a scale-free substrate for cognition and consciousness. *Neuroscience of Consciousness, 2021*(2), Article niab013. https://doi.org/10.1093/nc/niab013

18 Conclusion

Sheila L. Macrine, Jennifer M. B. Fugate, Arsen Abdulali, and Josie Hughes

Embodied intelligence (EI), a computational approach to understanding and designing intelligent behavior in physically embodied and situated agents, emphasizes the tight coupling between an agent's body, sensorimotor systems, and surroundings. It moves beyond the traditional, isolated view of the mind, illustrating the profound implications of embodied cognition (EC) (Barrett & Stout, 2024). The broad reconceptualization of EI in this volume invites us to fundamentally reconsider intelligence.

Through transdisciplinary investigations spanning both biological and artificial systems, the authors demonstrate how embodiment serves as the inextricable and foundational link between an entity's physical form and its cognitive capabilities (Clark, 2008). Expanding on the seminal insights of pioneers like Brooks (1991) and Pfeifer & Bongard (2006), this volume advances an inclusive and interdisciplinary view of EI, illustrating how an entity's physical form, or body, its cognitive apparatus, and its dynamic, ongoing interactions with the environment fundamentally shape its intelligent behavior. Said another way, intelligence is a dynamic and emergent phenomenon, arising from the continuous and reciprocal coupling of an agent with its specific and ever-changing environmental context. This perspective has evolved alongside advances in robotics, computational intelligence, neuroscience, and cognitive science, focusing on concepts like morphological computation (MC), sensorimotor coordination, and EC.

As a result, the volume explores how these embodied interactions, particularly in a world where the lines between biological and artificial systems are becoming increasingly blurred (Levin, 2022), influence our understanding of intelligence, consciousness, and even life itself. Ultimately, this exploration compels a deeper consideration of how the very nature of an intelligent agent, in its various forms of being and interacting—whether through a physical body, a virtual presence, or another potential instantiation of form—fundamentally underpins perception, understanding, and engagement with its environment. To lay the groundwork for this exploration, Part I delves into the fundamental principles and theoretical frameworks that underpin EI.

Theoretical Foundations and Philosophical Roots

Part I of this volume helps reshape our understanding of EI across diverse scientific, philosophical, and technological domains, beginning by establishing its critical philosophical and theoretical underpinnings, including beginning with an examination of parallels between human embodied cognition (HEC) and embodied artificial intelligence (EAI). This foundation highlights the similarities in how biological and artificial agents learn through environmental interaction and brings insightful attention to how disruptions to embodiment can significantly impact self-awareness that underscore the body's crucial role in our sense of self (Montupil et al., 2023; chapter 2).

Building on this emphasis on embodiment and the interconnectedness of body, environment, and cognition, this section critically explores key theoretical frameworks for understanding the mind, such as active inference and predictive processing (Clark, 2006; Friston, 2010; Friston et al., 2017; Di Paolo et al., 2024; chapters 3 and 4). These frameworks offer a unified account of perception, action, and cognition, bridging the historical divide between representationalist and dynamicist approaches, and emphasize that cognition is not an isolated process within the brain but is fundamentally shaped by the agent's interaction with its surroundings and its own agency.

A particularly influential example is the free energy principle (FEP) (Friston, 2010), which posits that both biological and artificial systems are fundamentally driven by an intrinsic imperative to minimize prediction error (or "free energy") through active engagement with their surrounding environment, validating and dynamically refining their internal models of the world in the process. The FEP provides an overarching principle that underlies the adaptive behaviors observed in embodied agents across a vast spectrum of complexity, from the simplest unicellular organisms exhibiting remarkable problem-solving capabilities (Levin, 2022) to the most sophisticated human beings and artificially intelligent systems navigating complex and dynamic environments.

Importantly, active inference, a key concept derived from the FEP, elucidates how organisms minimize prediction error through a continuous cycle of action and perception. Organisms are constantly engaged in a process of predicting their sensory inputs by minimizing the difference between predicted and actual sensory experience—a perspective that provides a powerful lens through which to understand how embodiment shapes cognition and behavior, with profound implications for the design and development of Artificial Intelligence and robotics.

Embodied AI: Applications, Ethics, and Future Directions

Transitioning from abstract theoretical considerations, part II gives significant and nuanced attention to the rapidly evolving field of EAI and the challenges and opportunities it presents. These chapters explore the intricate processes involved in creating intelligent agents grounded in cognitive architectures that emphasize physical embodiment and dynamic situatedness within complex, unpredictable environments (chapter 5). A key focus is on the inherent complexities of endowing artificial agents with genuine situated intelligence. To further address the complexities of endowing artificial agents with EAI's situated intelligence, this part advances an ecologically valid and biologically inspired methodology (chapter 6), drawing inspiration from the principles of self-organization and adaptation observed in natural biological systems.

While acknowledging the current state of EAI, its limitations, and its inherent challenges, the chapters also chart promising and innovative future directions. These directions aim to achieve the goal of creating more robust, adaptable, and genuine EAIs that are capable of context-aware and ethically sound interaction with the physical world and with human beings (chapter 7). The development of such agents inevitably raises significant ethical implications, which are carefully and rigorously considered (chapters 8 and 9), underscoring the paramount importance of responsible innovation, proactive ethical guidelines, and ongoing societal dialogue to govern this rapidly advancing field.

Embodied Intelligence: Emergent Properties and Diverse Embodiments

Broadening our perspective, part III challenges long-standing and often implicit anthropocentric biases that have historically and often unconsciously dominated the scientific study of intelligence. This section advocates for a broader, more inclusive, and ultimately more accurate conceptualization of intelligence that extends far beyond the traditional (and often limiting) confines of human cognition to one that emphasizes the critical and often underestimated roles of adaptation and morphology across a diverse range of biological and artificial beings (chapters 10 and 11). In doing so, this section reveals the surprising relationship between an agent's physical form (biological or artificial) and the emergence of complex behaviors and sophisticated problem-solving capabilities (chapter 12). Even simple physical systems, when thoughtfully coupled with appropriate and context-rich environmental interactions, can exhibit surprisingly sophisticated and remarkably adaptive forms of intelligence that challenge our preconceived notions of cognitive complexity (Levin, 2025). The amazing efficiency, flexibility, and adaptability observed in natural biological systems across vast timescales extends to both technological innovation and the future advancement of transformative potential of bioinspired approaches (Cangelosi et al., 2015; Kriegman et al., 2021).

From Morphology to Affect: The Social and Interactive Dimension

Part III focused on the physical and biological substrates of intelligence. Yet, true embodiment extends beyond mere morphology to encompass the internal regulatory processes and affective states that arise from maintaining a body in a complex world. Part IV examines the critical and often overlooked intersection of emotion, learning, and embodied interactions, recognizing these fundamental elements as integral and indispensable components in the development of more sophisticated, socially aware, and ultimately more human-like artificial agents (chapter 13). It underscores the fundamental importance of embodiment in facilitating the creation of believable, engaging, and contextually appropriate social interactions between humans and artificial agents, thereby paving the way for more natural, intuitive, and ultimately more effective forms of human-robot collaboration across a wide and expanding range of applications in fields such as mental health (chapter 14), education (chapter 15), and elder care.

In an era characterized by unprecedented and accelerating advancements in the field of AI, this part articulates a compelling and urgent argument for a proactive, collaborative, and thoughtful engagement to address the critical needs of endowing artificial agents with the capacity to understand and appropriately respond to subtle and complex social cues, to potentially experience and express emotions in meaningful and context-aware ways, and to engage in ethical decision-making that aligns with deeply held human values and evolving societal norms.

This forward-looking perspective suggests that the future of AI involves more than just computational power and algorithmic sophistication. It highlights the importance of creating artificial systems that can recognize and respond to human emotions. Furthermore, it envisions systems that may possess their own affect—the underlying feeling state that gives rise to specific emotions. The chapters explore the profound ways in which the design and rigorous analysis of AI systems are explicitly grounded in the principles of embodiment through which to gain a deeper and more nuanced understanding of the intricate mechanisms underlying human cognition and the complex processes of social learning.

Unifying Principles and Interdisciplinary Integration

Part V presents a diverse array of perspectives and emerging themes. The concluding chapters endeavor to identify potential unifying themes and principles of EI that arise from these discussions.

Bioinspired soft robotics (Zhao et al., 2024; chapter 16) exemplifies the power of learning from nature. By drawing inspiration from the flexibility, adaptability, and resilience of living organisms, this field is creating innovative robots capable of navigating complex environments, handling fragile objects with precision, and safely collaborating with humans in areas like health care and manufacturing. The physical form and material properties of these robots are key factors that significantly shape their overall capabilities and the nuanced nature of their interactions with the world. Consequently, this offers exciting and potentially transformative visions for future applications in areas like personalized health care, advanced manufacturing, and even deep-sea and space exploration.

The complex and inherently multifaceted questions surrounding EI necessitate a new approach. This section strongly emphasizes the critical and urgent need for a unified and comprehensive theoretical framework. This framework must thoughtfully prioritize the intricate interplay of an agent's intrinsic goals, its inherent capabilities, and its dynamic interactions with its surrounding environment across various systems. This is crucial to help move beyond the often-rigid and ultimately artificial distinctions.

The powerful concept of an agent's "cognitive light cone" (Levin, 2019, p. 9) represents the range of goals that an agent can understand and achieve within its environmental and physical constraints (chapter 17). This concept is presented as a particularly valuable and versatile tool for assessing the cognitive capabilities of a wide array of diverse entities. This thereby promotes a more inclusive, embodied, and ultimately more expansive understanding of the fundamental nature of intelligence. This understanding transcends traditional anthropocentric biases.

Further, Levin's (2022) thought-provoking and potentially revolutionary notion of "agential material" is presented. This notion posits that even seemingly simple biological systems, such as individual cells and tissues, can be viewed as active agents. These agents possess their own inherent goals and problem-solving abilities at a fundamental level of biological organization. This concept challenges long-held conventional categories. Moreover, it has profound and far-reaching implications for rapidly advancing fields such as regenerative medicine and synthetic biology. Additionally, it impacts the very engineering of novel lifeforms. Finally, it raises profound ethical questions about the fundamental nature of agency and autonomy in complex adaptive systems.

Bridging Biology, Bioengineering, and Technology

The integration of the biological sciences and the technological disciplines is presented as a central and overarching theme across the volume. This theme echoes the compelling and increasingly evident assertion that intelligence can manifest itself in a remarkably diverse array of forms and across a wide spectrum of underlying physical and computational substrates (Levin & Rouleau, 2024). This understanding moves decisively beyond simplistic and often limiting categorizations based on material composition. Furthermore, this volume actively explores the potentially synergistic and mutually beneficial relationship between the deep and time-tested insights gleaned from the rigorous study of biological systems over evolutionary timescales and the rapid—and often breathtaking—advancements occurring in the diverse fields of technological innovation.

The ongoing and accelerating advancements in bioengineering are blurring the traditional boundaries between humans, robots, biological-artificial hybrids, and purely mechanical machines. This blurring necessitates a fundamental evolution in our philosophical and psychological frameworks to accommodate these emerging hybrid forms of intelligent life, which challenge traditional definitions of life and cognition.

A groundbreaking and paradigm-shifting example is found in the research on Xenobots (Blackiston et al., 2021; Blackiston et al., 2023; Kriegman et al., 2020). These are the first living robots, ingeniously assembled from biological cells, where their morphology and behavior are determined by evolutionary algorithms. This serves as a compelling example of the profound implications of these remarkable advancements for our fundamental understanding of the very nature of life and intelligent behavior. It compels us to critically reconsider our own often-privileged place within the broader and increasingly complex spectrum of intelligent entities that inhabit our world and potentially beyond.

Similarly, in the realm of simulation, the principles of morphological adaptation are being explored through computer-generated physical agents, known as "Unimals" (Gupta et al., 2021), which are inspired by the efficient movement and adaptability of arthropods. Through interacting within the computer-simulated environment, they demonstrate how their interactions can directly shape the emergence of their intelligent behaviors, which are honed through sophisticated Deep Evolutionary Reinforcement Learning (DERL) algorithms (Gupta et al., 2021). By observing how different virtual morphologies lead to varying levels of perceptual ability and interactional finesse in navigating their world, this research provides valuable insights into the fundamental connection between embodiment and the development of intelligence, even in purely artificial systems.

Future Directions: Key Questions and Challenges

Looking toward the exciting and uncertain future of this rapidly evolving field, the volume thoughtfully charts several critical and promising directions for future research in EI.

The Nature of Form

A continuing challenge is whether the architecture of an agent fundamentally dictates how it perceives and interacts with its surrounding environment and whether it fully dictates the subjective experience of the world (Liu et al., 2025; Sun et al., 2024). This line of inquiry explores the profound question of whether the specific form or structure of an agent's body inherently shapes its perception, understanding, and interaction with the world. For instance, does the possession of a certain type of form/body or a specific type of internal physiology fundamentally constrain or afford certain ways of experiencing and acting in the world? Could radically different forms of embodiment, whether biological, artificial, or hybrid, lead to radically different forms of subjective experience, perception, and cognition?

Hardware Limitations

Furthermore, this volume acknowledges and addresses the significant challenges of developing reliable and adaptable EAI systems. These challenges are often daunting and encompass critical areas such as current hardware limitations. For example, there are inherent difficulties in ensuring sensor reliability and accuracy across diverse and unpredictable environmental conditions, and there are remaining challenges in processing and interpreting vast amounts of sensory data in real time. Another crucial need is improved model generalization across such unpredictable environments and the seamless integration of effective multimodal integration of multiple sensory inputs.

Overcoming these multifaceted challenges could pave the way for the creation of EAI systems with fundamentally different modes of perception and interaction, expanding the very definition of what it means to perceive, act, and interact with the world. This could lead to the development of artificial agents with entirely new sensory modalities, allowing them to perceive phenomena that are currently imperceptible to humans, such as magnetic fields, infrared radiation, or even higher-dimensional spatial relationships, potentially revolutionizing fields like scientific exploration, disaster response and mitigation, and advanced health care diagnostics and treatment. Despite these considerable challenges, researchers are making significant and rapid progress in developing innovative solutions, such as advanced algorithms and AI techniques. These advancements are enabling robots to perform increasingly complex tasks and interact with humans in new, exciting, and potentially transformative ways.

Context, Social-Awareness and Consciousness

Ongoing and intensive research efforts are actively striving to bridge these critical gaps through the development and refinement of a range of innovative technologies and approaches. These include neuromorphic computing, which seeks to mimic the brain's energy-efficient and parallel processing capabilities; brain-computer interfaces, which offer the potential for more seamless and intuitive interaction between humans and artificial agents; and biologically inspired AI design, which draws direct inspiration from the elegant and efficient solutions evolved by natural systems over millions of years (Pham et al., 2023).

These cutting-edge innovations, alongside continued progress in the field of explainable AI (Saeed & Omlin, 2023), are anticipated to significantly improve the adaptability, transparency, and ethical application of artificial neural networks (ANNs). This could foster the development of more human-like AI capabilities and enable the creation of more seamless and symbiotic human-AI systems that can collaborate effectively and ethically across a wide range of domains. This volume advocates thoughtful and proactive consideration of how the principles of EI can contribute positively to the development of safe, ethical, and beneficial interactions between humans and artificial agents, addressing the complex ethical implications of creating increasingly sophisticated and potentially autonomous embodied agents (Sun et al., 2024).

Furthermore, ongoing research into the neural correlates of consciousness and the development of artificial consciousness could potentially provide valuable insights into the design of truly sentient AI. However, this remains a highly debated and ethically complex area that demands careful and nuanced consideration from researchers, policymakers, and the broader public. Finally, the idea is emphasized that embodiment equips AI models with the crucial ability to interact meaningfully and effectively within our inherently human-centric world. This allows EAI to be more seamlessly integrated into our homes, workplaces, hospitals, and other social environments.

Beyond these technical hurdles lies a profound societal challenge: the ethics of coexistence. As we transition from using tools to collaborating with embodied agents—some of which may possess bio-hybrid substrates or rudimentary forms of sentience—we must develop new frameworks for moral patiency. The question is no longer just what these systems can do for us but how we can coexist within an ecosystem of diverse intelligences without reducing them to mere instruments.

This deeper understanding holds the potential for transformative applications in fields such as mental health care, where a more nuanced appreciation of the role of embodiment in cognition could lead to the development of new therapies for conditions involving disrupted embodiment, such as anxiety disorders, depression, and certain neurological conditions. It could also provide valuable insights into the fundamental nature of human consciousness, self-awareness, and subjective experience.

Toward a Unified Framework and Future Directions for Embodied Intelligence

To summarize, the contributors thoughtfully explore the remarkable breadth and depth of current research in the dynamic and rapidly expanding field of EI. This exploration fosters a deeper and more nuanced appreciation for the fundamental, and often subtle, ways in which embodiment shapes not only the core processes of cognition and perception but also the very subjective experience of being a situated and embodied agent within a complex and dynamic world. Consequently, it urges us to critically reconsider our long-held assumptions about the fundamental nature of consciousness and the elusive sense of self and inspires exciting new and interdisciplinary avenues for future scientific inquiry that promise to further illuminate the intertwined relationship between body, environment, and mind.

Ultimately, the convergence of the FEP and the physics of agential material offers a tantalizing path toward a unified science of mind. If intelligence is indeed the ability to navigate problem spaces—whether metabolic, anatomical, or spatial—then the distinction between "living" and "machine" becomes less about origin and more about the scale of competence. This unified view suggests that the engineering of future AI is not merely a task of coding but a task of orchestrating the natural tendency of matter to self-organize, minimize surprise, and expand its "cognitive light cone."

This volume explores the potential to develop core principles of a broadened EI. Such principles could significantly guide future EAI. The aim is to foster a far richer and more nuanced understanding of the crucial roles of an agent's physical form and its surrounding environment. A deeper grasp of these elements might even enable us to transcend the inherent limitations of our current three-dimensional human perspective, potentially opening avenues to explore cognitive architectures. Such architectures might operate within the mathematical frameworks of non-Euclidean geometries and under fundamentally different physical laws than those we experience.

This paradigm shift could pave the way for creating AI systems capable of inhabiting virtual or simulated environments. Such environments might be characterized by physical laws and dimensionality that are radically different from our own, raising profound and currently unanswered questions. These inquiries concern how vastly different conditions might shape the cognitive abilities of artificial systems, their very sense of self, and ultimately their potential for consciousness. Such considerations are fundamental inquiries at the intersection of AI, philosophy, and our understanding of the nature of intelligence itself.

Embodied Intelligence as a Science of Emergent Minds

This volume articulates a compelling and urgent call to fundamentally reimagine intelligence research across all entities. This reconceptualization boldly positions the study of intelligence as a vital and increasingly central science of emergent minds.

By fully acknowledging, embracing, and celebrating the inherent and often surprising potential for diverse forms of intelligence to arise from a multitude of different embodiments—whether those embodiments are composed of biological tissue, engineered materials, or computational code—the volume establishes the essential groundwork for a more nuanced, more ethically grounded, more intellectually rigorous, and fundamentally richer trajectory for the entire field of intelligence research, promising to unlock profound new insights into the nature of life, mind, and our place within the vast and awe-inspiring universe.

The realization of this new understanding of intelligence demands a deep integration of diverse fields, specifically a sharing and shaping of concepts toward epistemic alignment across disciplines. It's not simply about neuroscience, psychology, computer science, robotics, philosophy, and biology studying embodiment in isolation. It requires that these fields actively inform and shape each other's perspectives. The volume aims to create a framework that transcends the boundaries of any single discipline. Importantly, by making its insights accessible through open access, this volume fosters broader engagement and collaboration across disciplines.

Fostering a deeper, more integrated understanding of the core principles of EI is crucial. Actively promoting interdisciplinary collaboration and the open and accessible sharing of research findings across diverse fields is also essential for the development of a shared epistemic framework. Moreover, this approach can lead to a richer and more profound understanding of ourselves, our place in the universe, and the fundamental nature of intelligence itself, in all its diverse and wondrous manifestations.

To summarize, this volume represents a crucial step in an ongoing intellectual journey, seeking to lay the groundwork for a paradigm shift in how we approach the study of intelligence. It offers a compelling, transformative, and ethically grounded vision of a future where the increasing and inevitable convergences among biological and artificial systems not only expand the very possibilities of what we consider to be life itself but also fundamentally reshape our understanding of the multifaceted and often elusive phenomenon of intelligence.

By championing a transdisciplinary and embodied perspective, this volume opens up exciting new avenues for research and technological innovation. The questions and challenges explored in these chapters invite continued exploration and collaboration across disciplines, promising to fundamentally reshape our understanding of what it means to be intelligent, to perceive, to act, and to exist in the world, whether as a biological organism, an artificial agent, or something in between.

The pursuit of EI promises to be a defining endeavor of the twenty-first century, transforming our understanding of ourselves not as isolated minds but as part of a vast, emergent continuum of matter waking up to meet the world. By embracing this science of emergent minds, we step into a future where the boundaries between the born and the made dissolve, revealing a universe richer in agency and possibility than we ever imagined.

References

Barrett, L., & Stout D. (2024). Minds in movement: embodied cognition in the age of artificial intelligence. *Philosophical Transactions of the Royal Society B*, *379*, 1–9. https://doi.org/10.1098/rstb.2023.0144

Blackiston, D., Kriegman, S., Bongard, J., & Levin, M. (2023). Biological robots: Perspectives on an emerging interdisciplinary field. *Soft Robotics, 10*(4), 674–686. https://doi.org/10.1089/soro.2022.0142

Blackiston, D., Lederer, E., Kriegman, S., Garnier, S., Bongard, J., & Levin, M. (2021). A cellular platform for the development of synthetic living machines. *Science Robotics, 6*(52), eabf1571. https://doi.org/10.1126/scirobotics.abf1571

Brooks, R. A. (1991). Intelligence without representation. *Artificial Intelligence, 47*(1–3), 139–159.

Cangelosi, A., Bongard, J., Fischer, M. H., & Nolfi, S. (2015). Embodied intelligence. In J. Kacprzyk & W. Pedrycz (Eds.), *Springer handbook of computational intelligence* (pp. 697–714). Springer. https://doi.org/10.1007/978-3-662-43505-2_37

Clark, A. (2006). Sensorimotor skills and perception: Cognitive complexity and the sensorimotor frontier. *Aristotelian Society Supplementary, 80*, 43–65. https://doi.org/10.1111/j.1467-8349.2006.00136.x

Clark, A. (2008). *Supersizing the mind: Embodiment, action, and cognitive extension.* Oxford University Press. https://doi.org/10.1093/acprof:oso/9780195333213.001.0001

Di Paolo, L., White, B., Guénin-Carlut, A., Constant, A., & Clark, A. (2024). Active inference goes to school: The importance of active learning in the age of large language models. *Philosophical Transactions of the Royal Society B, 379*(1911), 20230148. https://doi.org/10.1098/rstb.2023.0148

Friston, K. J. (2010). The free-energy principle: A unified brain theory? *Nature Reviews Neuroscience, 112*, 127–138.

Friston, K. J., Lin, M., Frith, C. D., Pezzulo, G., Hobson, J. A., & Ondobaka, S. (2017). Active inference, curiosity and insight. *Neural Computation, 29*(10), 2633–2683. https://doi.org/10.1162/neco_a_00999

Gupta, A., Savarese, S., Ganguli, S., & Fei-Fei, L. (2021). Embodied intelligence via learning and evolution. *Nature Communications, 12*(1), 5721. https://doi.org/10.1038/s41467-021-25874-z

Kriegman, S., Blackiston, D., Levin, M., & Bongard, J. (2020). A scalable pipeline for designing reconfigurable organisms. *Proceedings of the National Academy of Sciences, 117*(4), 1853–1859. https://doi.org/10.1073/pnas.1910837117

Kriegman, S., Mohammadi Nasab, A., Blackiston, D., Steele, H., Levin, M., Kramer-Bottiglio, R., & Bongard, J. (2021). Scale-invariant robot behavior with fractals. *arXiv.* https://doi.org/10.48550/arXiv.2103.04876

Levin, M. (2019). The computational boundary of a "self": Developmental bioelectricity drives multicellularity and scale-free cognition. *Frontiers in Psychology, 10*, 9. https://doi.org/10.3389/fpsyg.2019.02688

Levin, M. (2022). The origin of body-maps and the evolution of cognitive closure. *Philosophical Transactions of the Royal Society B: Biological Sciences, 377*(1864), 20210398. https://doi.org/10.1098/rstb.2021.0398

Levin, M. (2025). The multiscale wisdom of the body: Collective intelligence as a tractable interface for next-generation biomedicine. *BioEssays, 47*(3), e202400196. https://doi.org/10.1002/bies.202400196

Levin, M., & Rouleau, N. (2024). Discussions of machine vs. living intelligence need more clarity. *Nature Machine Intelligence, 6,* 1408. https://doi.org/10.1038/s42256-024-00968-7

Liu, H., Guo, D., & Cangelosi, A. (2025). Embodied intelligence: A synergy of morphology, action, perception, and learning. *ACM Computer Survey, 57*(7). https://doi.org/10.1145/3717059

Montupil, J., Cardone, P., Staquet, C., Bonhomme, A., Defresne, A., Martial, C., Alnagger, N. L. N., Gosseries, O., & Bonhomme, V. (2023). The nature of consciousness in anaesthesia. *BJA Open, 8.* https://doi.org/10.1016/j.bjao.2023.100224

Pfeifer, R., & Bongard, J. (2006). *How the body shapes the way we think: A new view of intelligence.* MIT Press. https://mitpress.mit.edu/9780262537421/how-the-body-shapes-the-way-we-think/

Pham, M. D., D'Angiulli, A., Dehnavi, M. M., & Chhabra, R. (2023). From brain models to robotic embodied cognition: How does biological plausibility inform neuromorphic systems? *Brain Sciences, 13*(9), 1316. https://doi.org/10.3390/brainsci13091316

Saeed, A., & Omlin, C. W. (2023). Explainable artificial intelligence for autonomous driving: A comprehensive review. *IEEE Access, 11,* 102607–102641. https://doi.org/10.1109/ACCESS.2023.3317765

Sun, F., Chen, R., Ji, T., & Zhou, X. (2024). A comprehensive survey on embodied intelligence: Advancements, challenges, and future perspectives. *CAAI Artificial Intelligence Research, 3.* 9150042. https://doi.org/10.26599/AIR.2024.9150042

Zhao, Z., Wu, Q., Wang, J., Zhang, B., Zhong, C., & Zhilenkov, A. A. (2024). Exploring Embodied Intelligence in soft robotics: A review. *Biomimetics, 9*(4), 248. https://doi.org/10.3390/biomimetics9040248

Contributors

Arsen Abdulali, Ph.D., is assistant professor in robotics and AI in the School of Engineering and Material Science, Queen Mary University of London, United Kingdom.

Dor Abrahamson, Ph.D., is professor of cognition and development in Berkeley's Graduate School of Education, where he runs the UC Berkeley Embodied Design Research Laboratory.

Mahault Albarracin, Ph.D., is the director of research strategy at VERSES AI, Inc., and a scientist in cognitive computing, robotics, and social sciences.

Francesco Bianchini, Ph.D., is associate professor of logic and philosophy of science in the Department of Philosophy and Communication Studies at the University of Bologna, Italy.

Josh Bongard, Ph.D., is professor and runs the Morphology, Evolution & Cognition Laboratory at the University of Vermont.

Lola Cañamero, Ph.D., is professor and chair of neuroscience and robotics at the Neurocybernetics Team of the ETIS Lab, CY Cergy Paris University, France, and a visiting professor at the University of Hertfordshire, United Kingdom.

Andy Clark, Ph.D., is professor of cognitive philosophy and professor informatics, with an affiliation with the Sackler Centre for Consciousness Science at the University of Sussex in the United Kingdom.

Axel Constant, Ph.D., is a research fellow (engineering and design) at the School of Engineering and Informatics at the University of Sussex in the United Kingdom.

Gordana Dodig-Crnkovic, Ph.D., is professor of computer science at Chalmers University of Technology, Gothenburg, and a professor at the School of Innovation, Design and Engineering, Mälardalen University, Västerås, Sweden.

Dobromir Dotov, Ph.D., is an assistant professor in the Department of Biomechanics at the University of Nebraska Omaha.

Martin H. Fischer, Ph.D., is professor of cognitive sciences at the University of Potsdam in Germany. He directs an international master course and leads the Potsdam Embodied Cognition Group (PECoG).

Karl Friston, Ph.D., is professor at University College London, in the United Kingdom. He is a noted British neuroscientist and theoretician and an authority on brain imaging and theoretical neuroscience.

Tom Froese, PhD, is a professor at Okinawa Institute of Science and Technology Graduate University, Japan where he leads the Embodied Cognitive Science Unit.

Jennifer M. B. Fugate, Ph.D., is professor in psychology and directs the ABLE (affective, behavior, learning, and embodiment) lab for the able mind and body at Kansas City University in Missouri.

Shaun Gallagher, Ph.D., Hon. D.Phil., is professor and the Lillian and Morrie Moss Professor of Excellence in Philosophy at the University of Memphis, Tennessee, and a professorial fellow at the School of Liberal Arts, University of Wollongong Australia.

Conor Heins, Ph.D., is a research engineer lead at VERSES AI, Inc., and an affiliate of the Max Planck Institute of Animal Behavior, Department of Collective Behaviour, in Konstanz, Germany.

Laura B. Hensel, Ph.D., is a postdoc at the University of Glasgow, in Scotland.

Matej Hoffmann, Ph.D., is assistant professor at Czech Technical University in the Department of Cybernetics, in electrical engineering in Prague, Czech Republic.

Ryman Hashem, Ph.D., is a senior research fellow in biomedical engineering at the University of Oxford, in the United Kingdom.

David Howard, Ph.D., leads the Robotic Design and Interaction Group and is a principal research scientist in the Cyber Physical Systems program at CSIRO, Australia's national science body.

Josie Hughes, Ph.D., is assistant professor and head of the Computational Robot Design & Fabrication Lab CREATE-LAB in the Institute for Mechanical Engineering at l'École Polytechnique Fédérale de Lausanne (EPFL), in Switzerland.

Fumiya Iida, Ph.D., is professor of robotics and the director of Biologically Inspired Robotics Laboratory in the Department of Engineering, University of Cambridge, in the United Kingdom. He is also a fellow of Corpus Christi College in Texas.

Alex B. Kiefer, Ph.D., is a staff R&D engineer at VERSES AI, Inc., and an affiliate of the Centre for Consciousness and Contemplative Studies at Monash University, in Melbourne, Australia.

Michael Levin, Ph.D., is the Vannevar Bush Distinguished Professor at the Department of Biology, and director of the Allen Discovery Center at Tufts University in Massachusetts.

Sheila L. Macrine, Ph.D., is professor and a cognitive psychologist in the Department of STEM Education at the University of Massachusetts Dartmouth.

Stacy Marsella, Ph.D., is professor in the Khoury College of Computer Science with a joint appointment in psychology at Northeastern University in Massachusetts.

Alex Miklashevsky, Ph.D., is the scientific lead of the WiNoDa Knowledge Lab at the Natural History Museum in Berlin and a visiting scientist in the Potsdam Embodied Cognition Group (PECoG) at the University of Potsdam, Germany.

Zachary A. Pardos, Ph.D., is associate professor in the School of Education: Adaptive Learning and AI at University of California Berkeley.

Thomas Parr, Ph.D., is professor in the Nuffield Department of Clinical Neurosciences, University of Oxford, United Kingdom.

Shubhan Parag Patni is Ph.D. candidate at Czech Technical University in Prague.

Giovanni Pezzulo, Ph.D., is professor and researcher at the National Research Council of Italy, Institute of Cognitive Sciences and Technologies (ISTC-CNR) in Rome.

Lawrence Shapiro, Ph.D, is a professor in the Department of Philosophy at the University of Wisconsin.

Tobias Thejll-Madsen is- Ph.D. candidate at the University of Glasgow, Scotland.

Marcelo Worsley, Ph.D., is associate professor in computer science and learning sciences and directs the Technological Innovations for Inclusive Learning and Teaching (TIILT) lab at Northwestern University, Illinois.

Nutchanon Yongsatianchot, Ph.D., is a postdoctoral research associate at the Khoury College of Computer Sciences at Northeastern University, Massachusetts.

Carlotta Isabella Zona, Ph.D., is a research fellow at the Potsdam Embodied Cognition Group (PECoG), University of Potsdam, Germany.

Index

Page numbers in italics refer to figures and tables.

Publisher contact:
The MIT Press
Massachusetts Institute of Technology
77 Massachusetts Avenue, Cambridge, MA 02139
mitpress.mit.edu

EU Authorised Representative:
Easy Access System Europe, Mustamäe tee 50,
10621 Tallinn, Estonia
gpsr.requests@easproject.com

Printed by Integrated Books International,
United States of America